Unstoppable Owl

1) Understanding Negative Numbers

A **negative number** has a value that is less than zero. A minus symbol (−) is written directly in front of a negative number to show that it is negative. For instance, negative 5 is written as −5.

Examples:		
Negative 8	=	−8
Negative 23	=	−23
Negative 165	=	−165

Directions: *Fill in the blank spaces based on the information provided.* (Examples are shown.)

Ex) **Negative 2** = −2

Ex) *Negative 3* = −3

1) **Negative 73** =

2) **Negative 97** =

3) **Negative 44** =

4) **Negative 160** =

5) **Negative 9** =

6) **Negative 314** =

7) **Negative 68** =

8) **Negative 51** =

9) **Negative 1237** =

10) = −9

11) = −45

12) = −258

13) = −329

14) = −78

15) = −8

16) = −2041

17) = −546

18) = −50

Negative numbers are often used in real-life. If someone owes the bank $2,000, it could be said that this person's bank balance is −$2,000. And, if the temperature is 13 degrees below zero, the temperature could be described as being −13 degrees.

Directions: *Write the negative number that best describes each situation.*

19) **Kyle is playing a video game. He's been penalized many times and now has 25 points below zero. How many points does Kyle have?**

20) **Sarah doesn't have any money. In fact, she owes her sister $85. What is Sarah's net worth?**

2) Understanding Negative Numbers

On a number line, negative numbers are positioned to the left of zero and positive numbers are positioned to the right of zero.

Directions: *Fill in the blank spaces on the number line.* (An example is shown.)

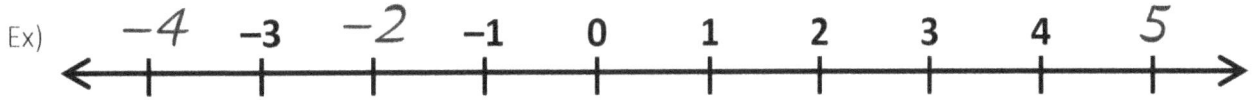

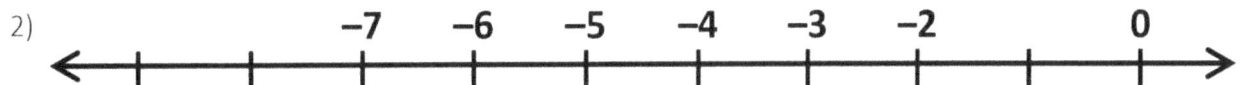

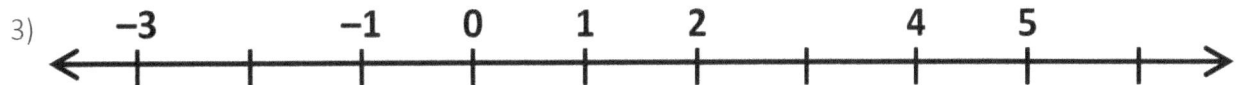

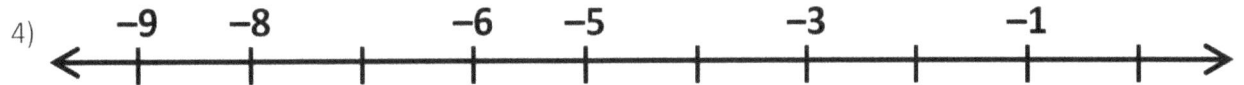

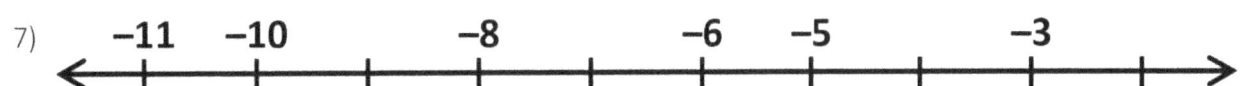

Ex)   −4   −3   −2   −1   0   1   2   3   4   5

1)   −6   −5   −4   −1   0   2   3

2)   −7   −6   −5   −4   −3   −2   0

3)   −3   −1   0   1   2   4   5

4)   −9   −8   −6   −5   −3   −1

5)   −2   0   1   3   5   7

6)   −7   −5   −2   −1   0   1   2

7)   −11   −10   −8   −6   −5   −3

3) Comparing Negative Numbers

Number lines can be used to visualize whether a negative number is **greater than** or **less than** another number. The number that is furthest left on a number line has the lesser value. Example 1 shows that −5 is greater than −12.

Example 1:   −5 $\boxed{>}$ −12

Example 2:   −35 $\boxed{<}$ −31

Directions: *Draw a <, >, or = symbol to compare each set of numbers.*

1) −2 $\square$ −8

2) −11 $\square$ −17

3) −50 $\square$ −21

4) −1 $\square$ 0

5) −73 $\square$ −65

6) −93 $\square$ −29

7) −44 $\square$ −13

8) −6 $\square$ −18

9) −31 $\square$ −72

10) −500 $\square$ −113

11) −200 $\square$ −201

12) −605 $\square$ −650

13) −104 $\square$ −302

14) −88 $\square$ −5

15) −9 $\square$ −4

16) −24 $\square$ −13

17) −71 $\square$ −80

18) −55 $\square$ −251

19) 0 $\square$ −78

20) −37 $\square$ −49

21) −815 $\square$ −627

Directions: *Fill in the blank spaces on the number line.*

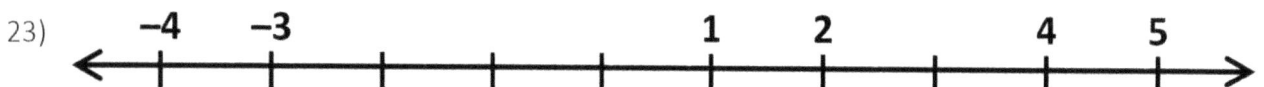

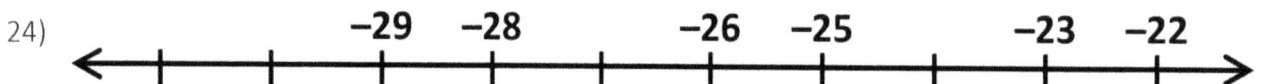

22) −58   −56   −54   −52 −51 −50

23) −4 −3   1   2   4   5

24) −29 −28   −26 −25   −23 −22

4) Comparing Negative Numbers

You've practiced comparing two numbers. Now, it's time to practice comparing three or more numbers. To do this, order the numbers from least to greatest.

Example 1 compares three numbers and example 2 compares four numbers.

Example 1:	–26, 5, –8
Answer:	*–26, –8, 5*
Example 2:	–7, 11, –13, 0
Answer:	*–13, –7, 0, 11*

Directions: *Rewrite each set of numbers so that they are ordered from **least to greatest**.*

1) –2, –8, –10

2) 0, –1, –5

3) –12, –9, –6

4) –3, 0, –13

5) 2, –4, –3

6) 0, 9, –9

7) –1, –7, 0

8) –6, 5, –13

9) –29, 18, –5

10) –4, –23, –8

11) –8, –1, –5, –3

12) –4, 0, 3, –7

13) 9, –9, –10, 10

14) 2, –6, –8, 0

15) –25, –17, –21, –11

16) –44, –29, –14, –32

17) –13, 6, –19, –20

18) –54, 60, –72, –81

19) 34, –41, –27, –18

20) –43, –75, –67, –59

21) –33, 41, –12, –20

22) –74, –80, 45, 29

23) 0, –39, –40, –59

24) 52, –54, –58, –43

25) –82, 61, –40, –94

26) –78, 84, –300, –99

27) –80, –15, 0, –49

28) –100, 99, –73, –81

29) –88, –90, –76, –91

30) –33, –27, –51, –45

5) Absolute Value

Absolute value is how far away a number is from zero. The symbol for absolute value is called a **modulus** (some people call it a **mod**). A modulus looks like two bars that are placed on either side of the number that you should find the absolute value for.

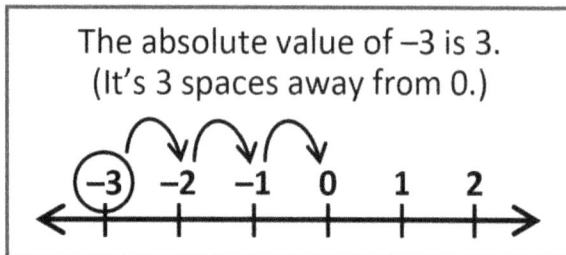

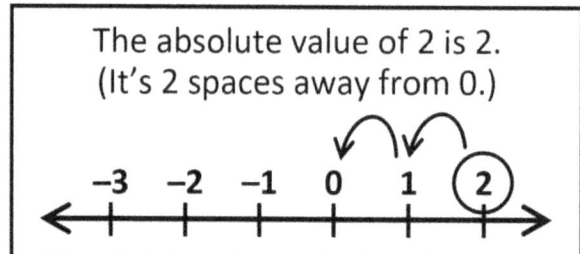

The absolute value of –3 is 3. (It's 3 spaces away from 0.)

–3   –2   –1   0   1   2

The absolute value of 2 is 2. (It's 2 spaces away from 0.)

–3   –2   –1   0   1   2

The absolute value of a number is never negative. It will either be a positive number or 0. (The only number with an absolute value of 0 is 0. All other numbers will have a positive absolute value.)

Notice that a negative number and a positive number can both have the same absolute value. (–6 and 6 are both 6 spaces away from 0. So, they each have an absolute value of 6.)

Examples:

$| -6 | = 6$

$| 6 | = 6$

$| -54 | = 54$

$| 18 | = 18$

$| -276 | = 276$

Directions: *Write the absolute value of each number.*

1) $| -2 | =$

2) $| -9 | =$

3) $| -15 | =$

4) $| -27 | =$

5) $| 19 | =$

6) $| -32 | =$

7) $| -11 | =$

8) $| 11 | =$

9) $| -5 | =$

10) $| 29 | =$

11) $| -19 | =$

12) $| -25 | =$

13) $| -64 | =$

14) $| -153 | =$

15) $| -241 | =$

16) $| 85 | =$

17) $| -98 | =$

18) $| 40 | =$

19) $| -66 | =$

20) $| -321 | =$

21) $| -8 | =$

22) $| -33 | =$

23) $| 64 | =$

24) $| -129 | =$

25) $| 270 | =$

26) $| -815 | =$

27) $| -362 | =$

28) $| 714 | =$

29) $| -608 | =$

30) $| -1000 | =$

6) Adding Negative Numbers

Adding with negative numbers takes a little thought. Using a number line to visualize the addition problem can help. Start with the first number of the addition problem and find that number on the number line. Then, look at the number that is being added.

- If the number being added is a **positive**, move that many spaces **to the** <u>right</u>.

- If the number being added is a **negative**, move that many spaces **to the** <u>left</u>.

Example 1: $-7 + 2 =$ **Step 1:** Start at -7. **Step 2:** Move 2 spaces <u>right</u>.) -8 (-7) -6 -5 **Answer:** $-7 + 2 = -5$	**Example 2:** $1 + -3 =$ **Step 1:** Start at 1. **Step 2:** Move 3 spaces <u>left</u>.) -2 -1 0 (1) **Answer:** $1 + -3 = -2$	**More Examples:** $10 + -4 = 6$ $-2 + -1 = -3$ $-3 + 5 = 2$ $0 + -6 = -6$ $2 + -2 = 0$

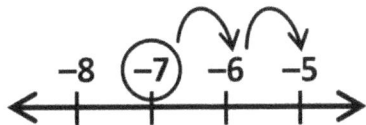

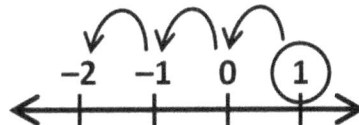

Directions: *Solve each problem. Draw a number line on a separate sheet of paper if it helps.*

1) $-9 + -2 =$

2) $-8 + 3 =$

3) $0 + -7 =$

4) $-6 + 0 =$

5) $-10 + -4 =$

6) $-12 + -3 =$

7) $5 + -2 =$

8) $-9 + 9 =$

9) $-11 + 10 =$

10) $15 + -7 =$

11) $-13 + 6 =$

12) $5 + -8 =$

13) $-14 + 7 =$

14) $-20 + 10 =$

15) $9 + -4 =$

16) $-16 + 1 =$

17) $6 + -8 =$

18) $-10 + -3 =$

19) $-5 + 19 =$

20) $18 + -9 =$

21) $-6 + 0 =$

22) $-7 + 15 =$

23) $-19 + -2 =$

24) $14 + -4 =$

25) $-11 + -8 =$

26) $13 + 2 =$

27) $5 + -6 =$

28) $-17 + -8 =$

29) $-20 + 20 =$

30) $-12 + 0 =$

7) Adding Negative Numbers

Did you notice any patterns when completing the addition problems in the previous section? Did you know that…

- When you add **two negative numbers,** the answer is always **negative.**

- When you add **two positive numbers,** the answer is always **positive.**

It's addition problems that contain a positive number and a negative number that can be confusing. Sometimes, their answer is positive and sometimes, it's negative.

The good news is **absolute values** can be used to determine whether the answer will be positive or negative. (That's right, you learned about absolute value for a reason.) Absolute values can even be used to solve the problem.

Find the absolute value of both numbers in the addition problem. Subtract from the larger absolute value with the lesser absolute value. Then, the number with the greatest absolute value will determine whether the answer will be negative or positive.

- If the number with the greatest absolute value is negative, the answer will be **negative.**

- If the number with the greatest absolute value is positive, the answer will be **positive.**

Example: $-7 + 5 =$

Step 1: Find both absolute values. ($| -7 | = 7$ and $| 5 | = 5$)
Step 2: Subtract from the larger absolute value. ($7 - 5 = 2$)
Step 3: Find the largest absolute value. (-7 has the larger absolute value.)
Step 4: Is it positive or negative? (-7 is negative. So, the answer will be **negative.**)

Answer: $-7 + 5 = -2$

Directions: *Solve each problem.*

1) $-42 + -8 =$

2) $-28 + -6 =$

3) $-4 + -34 =$

4) $-7 + -25 =$

5) $36 + -9 =$

6) $-55 + 15 =$

7) $-29 + 2 =$

8) $-11 + 26 =$

9) $-44 + -10 =$

10) $-23 + 15 =$

11) $-39 + -6 =$

12) $-28 + 8 =$

13) $-15 + -4 =$

14) $66 + -6 =$

15) $30 + -5 =$

16) $-48 + -1 =$

17) $24 + -7 =$

18) $-33 + 9 =$

19) $-40 + -4 =$

20) $-61 + 1 =$

21) $-8 + -88 =$

8) Adding Negative Numbers

You've practiced adding positive and negative numbers. Now, it's time to apply those skills to larger (and smaller) numbers. Take your time. Think about each problem and its answer.

Directions: *Solve each problem.*

1) −762 + −22 =

2) −102 + 25 =

3) −11 + −204 =

4) −10 + −314 =

5) 225 + −25 =

6) −522 + −26 =

7) 140 + −28 =

8) 201 + −15 =

9) −145 + −19 =

10) 356 + −56 =

11) −49 + −210 =

12) 217 + −25 =

13) −48 + −150 =

14) 387 + −80 =

15) −50 + −50 =

16) −23 + −98 =

17) 65 + −47 =

18) −500 + 200 =

19) −66 + −14 =

20) 85 + −33 =

Directions: *Write an addition equation for this word problem. Then, calculate the answer.*

21) Nancy is reading the bank statement for her business. The business checking account has a balance of $ −270. The business savings account has a balance of $1,100. How much money do these two accounts have combined?

9) Subtracting Negative Numbers

Subtracting with negative numbers is like adding with negative numbers. The steps are the same but reversed. Start with the first number of the subtraction problem and find that number on the number line. Then, look at the number that is being subtracted.

- If the number being subtracted is a **positive**, move that many spaces **to the left**.

- If the number being subtracted is a **negative**, move that many spaces **to the right**.

Example 1: $-2 - 1 =$ **Step 1:** Start at -2. **Step 2:** Move 1 spaces <u>left</u>.) $-3 \;\; (-2) \;\; -1 \;\;\; 0$ **Answer:** $-2 - 1 = -3$	**Example 2:** $-1 - -3 =$ **Step 1:** Start at -1. **Step 2:** Move 3 spaces <u>right</u>.) $(-1) \;\;\; 0 \;\;\; 1 \;\;\; 2$ **Answer:** $-1 - -3 = 2$	**More Examples:** $-8 - 2 = -10$ $4 - -5 = 9$ $-7 - -3 = 4$ $0 - -7 = 7$ $1 - 4 = -3$

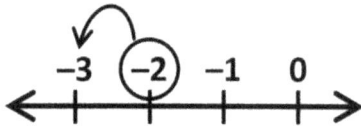

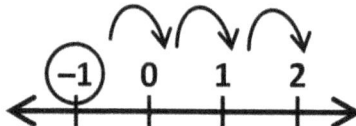

Directions: *Solve each problem. Draw a number line on a separate sheet of paper if it helps.*

1) $-5 - -4 =$

2) $-9 - -1 =$

3) $0 - -8 =$

4) $-7 - 0 =$

5) $-11 - -2 =$

6) $-6 - -5 =$

7) $-9 - 7 =$

8) $2 - -6 =$

9) $-12 - 6 =$

10) $10 - -5 =$

11) $-14 - 4 =$

12) $-7 - 6 =$

Subtraction symbols and negative symbols look similar. To avoid confusion, problems that contain negative numbers often include parentheses to help make the negative numbers more distinct. For instance, $5 - -2 = 7$ can be written as $5 - (-2) = 7$.

13) $6 - (-9) =$

14) $(-1) - 0 =$

15) $(-7) - (-3) =$

16) $8 - 8 =$

17) $(-10) - (-5) =$

18) $4 - (-6) =$

19) $(-14) - 7 =$

20) $(-3) - (-2) =$

21) $11 - 7 =$

22) $(-5) - 12 =$

23) $(-13) - (-9) =$

24) $19 - (-1) =$

10) Subtracting Negative Numbers

Directions: *Solve each problem.*

1) $(-12) - (-6) =$

2) $7 - (-5) =$

3) $(-8) - 13 =$

4) $(-16) - (-2) =$

5) $20 - 9 =$

6) $(-19) - 0 =$

7) $(-3) - (-18) =$

8) $(-1) - (-15) =$

9) $72 - (-6) =$

10) $34 - (-2) =$

11) $84 - (-4) =$

12) $(-29) - 9 =$

13) $(-14) - (-14) =$

14) $(-75) - 5 =$

15) $8 - (-45) =$

16) $(-6) - (-74) =$

17) $0 - (-88) =$

18) $(-50) - 4 =$

19) $(-73) - (-3) =$

20) $(-48) - 8 =$

21) $13 - 12 =$

22) $(-75) - 2 =$

23) $(-1) - (-91) =$

24) $27 - (-3) =$

25) $(-44) - 6 =$

26) $10 - (-33) =$

27) $(-57) - 8 =$

28) $(-94) - (-5) =$

29) $45 - 7 =$

30) $(-87) - 0 =$

Take your time and solve subtraction problems that have larger (and smaller) numbers.

31) $(-500) - 165 =$

32) $(-245) - (-125) =$

33) $(-310) - 24 =$

34) $-96 - (-147) =$

35) $(-507) - (-112) =$

36) $431 - (-37) =$

37) $(-66) - 1{,}016 =$

38) $(-217) - (-110) =$

39) $2{,}531 - 30 =$

40) $89 - (-901) =$

11) Addition & Subtraction Mixed Review

Directions: *Solve each problem.*

1) $(-2) + 0 =$

2) $0 - (-10) =$

3) $(-1) + 15 =$

4) $(-10) - (-6) =$

5) $19 - 5 =$

6) $(-8) + 3 =$

7) $(-2) + (-11) =$

8) $17 - (-7) =$

9) $21 - (-5) =$

10) $(-34) + (-4) =$

11) $(-50) - 9 =$

12) $7 + (-17) =$

13) $(-65) - 10 =$

14) $(-78) - (-2) =$

15) $0 + (-66) =$

16) $(-78) - 0 =$

17) $12 - 47 =$

18) $(-40) - 8 =$

19) $(-62) + (-3) =$

20) $9 + (-72) =$

21) $(-56) - 7 =$

22) $16 + (-8) =$

23) $(-42) - 6 =$

24) $(-20) - (-2) =$

Think you've mastered addition and subtraction? Move on to the next section.
Need a little more practice? Take a break. Then, complete the problems below.

25) $(-11) - (-9) =$

26) $(-71) + 0 =$

27) $24 + 10 =$

28) $67 - (-3) =$

29) $(-29) + 28 =$

30) $40 - (-7) =$

31) $(-12) - 12 =$

32) $(-32) + (-9) =$

33) $(-22) - 8 =$

34) $30 - (-7) =$

35) $(-65) + 40 =$

36) $(-80) - (-2) =$

37) $55 + 5 =$

38) $(-15) + 14 =$

39) $(-35) - (-6) =$

40) $20 + (-13) =$

41) $25 + 25 =$

42) $(-64) - (-10) =$

43) $(-102) - 5 =$

44) $200 + (-20) =$

45) $(-88) + 18 =$

46) $(-74) - (-9) =$

47) $16 + (-56) =$

48) $(-35) - 25 =$

12) Multiplying Negative Numbers

Multiplying negative numbers is easy, *as long as you remember a few rules.*

- If the numbers being multiplied are **both positive,** the answer will be **positive.**   **(+ + = +)**

- If the numbers being multiplied are **both negative,** the answer will be **positive.**   **(− − = +)**

- When multiplying **a positive and a negative,** the answer will be **negative.**   **(+ − = −)**

- When multiplying **a negative and a positive,** the answer will be **negative.**   **(− + = −)**

Example 1: $-2 \times 4 =$	Example 2: $-5 \times -3 =$	**More Examples:**
(negative × positive = negative)	(negative × negative = positive)	$4 \times -3 = -12$
Answer: $-2 \times 4 = -8$	Answer: $-5 \times -3 = 15$	$-2 \times -7 = 14$
		$5 \times 2 = 10$

Directions: *Solve each problem.*

1) $(-3) \times (-4) =$

2) $6 \times (-8) =$

3) $(-5) \times 4 =$

4) $(-6) \times (-7) =$

5) $9 \times 9 =$

6) $(-7) \times 8 =$

7) $(-4) \times (-2) =$

8) $(0) \times (-9) =$

9) $4 \times (-6) =$

10) $5 \times (-8) =$

11) $12 \times (-1) =$

12) $(-10) \times 1 =$

13) $(-6) \times (-5) =$

14) $(-7) \times 3 =$

15) $2 \times (-9) =$

16) $(-4) \times (-8) =$

17) $9 \times (-3) =$

18) $(-2) \times 6 =$

19) $(-4) \times (-4) =$

20) $(-6) \times 8 =$

21) $7 \times 7 =$

22) $(-1) \times 11 =$

23) $(-9) \times (-8) =$

24) $7 \times (-2) =$

25) $(-12) \times 2 =$

26) $2 \times (-11) =$

27) $(-5) \times 11 =$

28) $(-6) \times (-10) =$

29) $6 \times 6 =$

30) $(-12) \times 0 =$

13) Multiplying Negative Numbers

Directions: *Solve each problem.*

1) $(-8) \times (-6) =$

2) $12 \times (-3) =$

3) $(-4) \times 11 =$

4) $2 \times (-10) =$

5) $(-5) \times 9 =$

6) $(-4) \times (-2) =$

7) $12 \times 0 =$

8) $(-6) \times 6 =$

9) $3 \times (-7) =$

10) $(-5) \times 4 =$

11) $(-6) \times (-7) =$

12) $3 \times 9 =$

13) $(-2) \times 11 =$

14) $(-12) \times (-5) =$

15) $(-9) \times (-11) =$

16) $10 \times (-1) =$

17) $(-11) \times (-3) =$

18) $(-6) \times 12 =$

19) $0 \times (-2) =$

20) $(-9) \times (-8) =$

21) $3 \times (-4) =$

22) $(-9) \times 4 =$

23) $(-8) \times (-7) =$

24) $(-2) \times 10 =$

Think you've mastered multiplication? Move on to the next section.
Need a little more practice? Take a break. Then, complete the problems below.

---

25) $(-10) \times 10 =$

26) $(-11) \times (-4) =$

27) $(-5) \times 9 =$

28) $2 \times (-8) =$

29) $(-3) \times (-6) =$

30) $5 \times (-5) =$

31) $(-1) \times 6 =$

32) $(-6) \times (-7) =$

33) $12 \times (-6) =$

34) $(-3) \times 10 =$

35) $(-9) \times (-6) =$

36) $5 \times 11 =$

37) $(-8) \times 3 =$

38) $(-5) \times (-12) =$

39) $(-6) \times (-8) =$

40) $4 \times (-2) =$

41) $0 \times 1 =$

42) $(-1) \times 9 =$

43) $(-2) \times (-8) =$

44) $7 \times (-3) =$

45) $(-4) \times 6 =$

46) $7 \times (-7) =$

47) $(-9) \times 8 =$

48) $11 \times 12 =$

14) Dividing Negative Numbers

Know how to multiply negative numbers? Well, the rules are the same for division.

- If the numbers being divided are **both positive,** the answer will be **positive.**   **(+ + = +)**

- If the numbers being divided are **both negative,** the answer will be **positive.** **(− − = +)**

- When dividing **a positive and a negative,** the answer will be **negative.**   **(+ − = −)**

- When dividing **a negative and a positive,** the answer will be **negative.**   **(− + = −)**

Example 1: $-12 \div 4 =$	Example 2: $-6 \div -3 =$	More Examples:
(negative ÷ positive = negative)	(negative ÷ negative = positive)	$8 \div -2 = -4$
Answer: $-12 \div 4 = -3$	**Answer:** $-6 \div -3 = 2$	$-14 \div -7 = 2$
		$10 \div 2 = 5$

Directions: *Solve each problem.*

1) $1 \div (-1) =$

2) $(-8) \div 2 =$

3) $(-9) \div (-3) =$

4) $16 \div 4 =$

5) $(-25) \div 5 =$

6) $42 \div (-6) =$

7) $(-6) \div 2 =$

8) $(-14) \div (-2) =$

9) $(-16) \div 8 =$

10) $81 \div (-9) =$

11) $(-11) \div (-1) =$

12) $12 \div (-12) =$

13) $(-18) \div 3 =$

14) $(-60) \div (-5) =$

15) $(-121) \div 11 =$

16) $8 \div 4 =$

17) $(-56) \div 7 =$

18) $(-32) \div (-8) =$

19) $24 \div (-6) =$

20) $(-49) \div 7 =$

21) $(-12) \div 2 =$

22) $(-20) \div (-2) =$

23) $(-6) \div (-6) =$

24) $35 \div (-5) =$

25) $56 \div (-8) =$

26) $(-21) \div (-3) =$

27) $36 \div (-3) =$

28) $(-30) \div 5 =$

29) $(-72) \div (-8) =$

30) $64 \div 8 =$

15) Dividing Negative Numbers

Directions: *Solve each problem.*

1) $(-20) \div 4 =$

2) $(-45) \div (-5) =$

3) $24 \div (-6) =$

4) $(-63) \div 9 =$

5) $40 \div (-8) =$

6) $(-90) \div 9 =$

7) $(-100) \div (-10) =$

8) $22 \div 11 =$

9) $(-18) \div (-2) =$

10) $15 \div (-3) =$

11) $(-28) \div 4 =$

12) $(-50) \div (-5) =$

13) $66 \div 6 =$

14) $(-21) \div 3 =$

15) $(-56) \div (-7) =$

16) $(-72) \div (-9) =$

17) $(-60) \div (-12) =$

18) $(-55) \div 5 =$

19) $70 \div (-7) =$

20) $(-9) \div (-1) =$

21) $16 \div (-8) =$

22) $(-7) \div 1 =$

23) $(-30) \div (-6) =$

24) $(-40) \div 8 =$

Think you've mastered division? Move to the next section.
Need a little more practice? Take a break. Then, complete the problems below.

---

25) $2 \div (-1) =$

26) $(-15) \div 3 =$

27) $(-21) \div (-7) =$

28) $49 \div 7 =$

29) $(-90) \div (-10) =$

30) $33 \div (-3) =$

31) $(-72) \div 8 =$

32) $(-45) \div (-5) =$

33) $20 \div (-4) =$

34) $(-100) \div 10 =$

35) $(-144) \div (-12) =$

36) $132 \div 11 =$

37) $(-10) \div 10 =$

38) $(-60) \div (-12) =$

39) $(-4) \div (-1) =$

40) $55 \div (-5) =$

41) $(-121) \div (-11) =$

42) $144 \div (-12) =$

43) $(-81) \div 9 =$

44) $66 \div (-6) =$

45) $(-30) \div 3 =$

46) $(-12) \div (-1) =$

47) $55 \div 5 =$

48) $(-80) \div 8 =$

16) Mixed Review

You've practiced each of the 4 operations. Now, it's time to practice keeping these skills straight. Take your time. Think carefully. Don't confuse the steps needed for each operation.

Directions: *Solve each problem.*

1) $(-10) \times (-5) =$

2) $(-62) - 8 =$

3) $7 + (-12) =$

4) $(-18) \div 3 =$

5) $(-4) \times 9 =$

6) $5 - (-11) =$

7) $(-70) \div (-7) =$

8) $(-65) + 40 =$

9) $(-30) - (-2) =$

10) $(-7) + (-8) =$

11) $(-5) - 9 =$

12) $2 \times (-8) =$

13) $(-3) \times (-6) =$

14) $20 \div (-5) =$

15) $(-4) \times 6 =$

16) $(-6) + (-7) =$

17) $3 - (-7) =$

18) $(-5) \times 4 =$

19) $(42) \div (-7) =$

20) $3 + 9 =$

21) $(-2) - 11 =$

22) $(-12) + (-5) =$

23) $(-9) \times (-11) =$

24) $10 \times (-1) =$

25) $(-24) \div 3 =$

26) $(-3) + 10 =$

27) $(-54) \div (-6) =$

28) $5 - (-11) =$

29) $(-8) \times 3 =$

30) $(-55) \div (-11) =$

31) $(-6) + (-8) =$

32) $18 \div (-6) =$

33) $(-11) \times (-3) =$

34) $(-6) \times 12 =$

35) $0 - (-2) =$

36) $(-9) - (-8) =$

37) $0 \times (-16) =$

38) $(-9) + 4 =$

39) $(-8) - (-7) =$

40) $(-56) \div (-7) =$

41) $0 - 1 =$

42) $(-1) \times 9 =$

43) $(-2) + (-8) =$

44) $25 - (-3) =$

45) $(-4) + 6 =$

46) $7 \times (-7) =$

47) $(-9) + 8 =$

48) $(-64) \div 8 =$

17) Mixed Review

Directions: *Solve each problem.*

1) $(-8) + (-6) =$

2) $10 \times (-5) =$

3) $(-4) \times 11 =$

4) $2 - (-10) =$

5) $(-45) \div 5 =$

6) $(-33) + (-6) =$

7) $12 - 0 =$

8) $(-30) \div 10 =$

9) $(-3) - (-7) =$

10) $(-2) \times 2 =$

11) $(-85) + (-33) =$

12) $72 \div 9 =$

13) $(-48) - 12 =$

14) $(-8) + (-4) =$

15) $(6) \times (-6) =$

16) $1 \div (-1) =$

17) $(-9) \div (-9) =$

18) $(-6) \times 10 =$

19) $0 \times (-8) =$

20) $9 - (-11) =$

21) $36 \div 6 =$

22) $(-3) + 7 =$

23) $26 - 17 =$

24) $(-12) + 10 =$

*Take a break, but don't skip this section. It's easy to mix up the steps for the operations. Focus on **accuracy**. The goal is to consistently complete these problems correctly.*

25) $(-10) + 10 =$

26) $(-11) - (-4) =$

27) $(-8) \times 7 =$

28) $12 \div (-3) =$

29) $(-14) \div (-2) =$

30) $8 \times (-8) =$

31) $(-10) + 5 =$

32) $(-9) - (-7) =$

33) $35 \div 5 =$

34) $(-9) \times 4 =$

35) $(-8) + (-2) =$

36) $23 - 11 =$

37) $(-9) + 4 =$

38) $(-4) \times (-10) =$

39) $(-7) \div (-1) =$

40) $5 - (-3) =$

41) $9 \times 0 =$

42) $(-11) \times 4 =$

43) $(-64) + (-10) =$

44) $(-78) - (-12) =$

45) $60 \div (-10) =$

46) $11 + (-8) =$

47) $(-49) - 11 =$

48) $5 + (-8) =$

18) Mixed Review

Directions: *Solve each problem.*

1) (−9) ÷ (−3) =

2) 22 × (−2) =

3) (−42) + 11 =

4) (−56) − 16 =

5) (−8) × 2 =

6) (−65) + (−40) =

7) 12 ÷ 1 =

8) (−200) − 74 =

9) 4 × (−2) =

10) (−6) × 5 =

11) (−7) + (−8) =

12) 25 − 10 =

13) (−33) ÷ 11 =

14) (−12) ÷ (−6) =

15) (−22) + (−22) =

16) 11 − (−1) =

17) (−6) + (−4) =

18) (−3) × 12 =

19) 0 × (−5) =

20) (−8) + (−9) =

21) 15 ÷ (−5) =

22) (−30) − 7 =

23) (−2) ÷ (−1) =

24) (−3) − (−8) =

*Take a break, but don't skip this section. It's easy to mix up the steps for the operations. Focus on **accuracy**. The goal is to consistently complete these problems correctly.*

25) (−9) × 8 =

26) (−11) × (−7) =

27) (−17) − 9 =

28) 47 + (−4) =

29) (−5) × (−9) =

30) 4 ÷ (−4) =

31) (−22) + 14 =

32) (−49) ÷ (−7) =

33) 10 + (−4) =

34) (−6) × 10 =

35) (−8) × (−9) =

36) 9 + 11 =

37) (−3) × 3 =

38) (−33) − (−16) =

39) (−18) + 2 =

40) 16 ÷ (−2) =

41) (−81) ÷ 9 =

42) (−9) × 7 =

43) (−5) × (−4) =

44) 12 ÷ (−3) =

45) (−9) + 0 =

46) 10 × (−2) =

47) (−25) − 25 =

48) 11 × 9 =

19) Adding Negative Numbers

Knowing *how* to use the operations with negative numbers is great. But knowing *when* to use each of the operations is important too. The following word problems simulate real-world scenarios. Look for keywords to help you decide which operation to use. Then, use your skills to solve each problem.

1) Judy is exploring a cave. She reached a depth of −56 feet. But as she followed the cave's tunnel further, she rose 8 feet. What is Judy's depth now?

2) It was −7 degrees outside. Then, a storm came and the temperature dropped 11 degrees. What is the temperature now?

3) Tyler, Sarah, Steve, and Rachel each have a −$8 tab at the golf club. How much money is on their tabs combined?

4) Alex had $60 in his checking account. Then, he wrote a check for $200. Once the check clears, how much money will he have in the account?

5) It was −8 degrees outside. 4 hours later, the temperature had risen 12 degrees. What is the temperature now?

6) Amy had a balance of−$300 in her savings account. Then, she deposited $500 into the account. How much money does Amy have in her account now?

7) David had 9 overdraft fees added to his account last year. If each overdraft fee the bank added was −$8, what was the total amount of overdraft fees added to his account?

8) A whale was at a depth of −30 meters. Then, it began to rise 2 meters per second. At this rate, how long will it take the whale to reach the surface?

20) Adding Negative Numbers

1) Ryan's bank account balance was −$45. Then, he deposited $50 into the account. How much money is in his account now?

2) A submarine was at a depth of −250 meters. Then, it dove 25 meters lower. What is the submarine's depth now?

3) Laura had $8 in her bank account. Then, her bank added a −$15 fee to the account. How much money does Laura have in her account now?

4) Jack, Susie, and Ben own an ice cream shop. Their business checking account currently has a balance of −$12,000. How much money is that per person?

5) Melissa is playing a board game. She had −80 points. Then, she drew a card that gave her 100 more points. How many points does she have now?

6) It was −32 degrees outside in the morning. By the afternoon, it had risen 20 degrees. What is the afternoon temperature?

7) Claire has $200 in her savings account. But she noticed the bank had added a −$10 fee to her account last week. She called the bank and complained. The bank agreed to subtract this fee from her account. How much money is in Claire's savings account now?

8) If the value of a truck depreciates by −$2,000 each year, how much will the truck's value depreciate by after 4 years?

21) Exponents

Repeated multiplication can be represented with an **exponent**. The **base** is the number being multiplied, and the **exponent** (or **power**) is how many times it is multiplied. The exponent is usually written as a smaller number to the upper right of the base number.

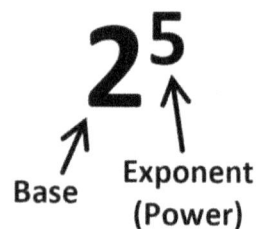

$$2^5$$

Base   Exponent (Power)

Word Form:	Exponential Form:	Expanded Form:	Value:
Two to the fifth power.	$= 2^5$	$= 2 \times 2 \times 2 \times 2 \times 2$	$= 32$
Four to the third power.	$= 4^3$	$= 4 \times 4 \times 4$	$= 64$
Nine to the second power.	$= 9^2$	$= 9 \times 9$	$= 81$
Three to the sixth power.	$= 3^6$	$= 3 \times 3 \times 3 \times 3 \times 3 \times 3$	$= 729$

Directions: *Fill in the blank spaces.*

	Word Form:	Exponential Form:	Expanded Form:
Ex)	Five to the fourth power.	$= 5^4$	$= 5 \times 5 \times 5 \times 5$
1)	Eight to the second power.	$=$	$=$
2)	Nine to the seventh power.	$=$	$=$
3)	Two to the fourth power.	$=$	$=$
4)	Ten to the fifth power.	$=$	$=$
5)		$= 1^8$	$=$
6)		$= 4^4$	$=$
7)		$= 11^3$	$=$
8)		$= 7^5$	$=$
9)		$=$	$= 2 \times 2$
10)		$=$	$= 12 \times 12 \times 12 \times 12 \times 12 \times 12$
11)		$=$	$= 3 \times 3 \times 3 \times 3 \times 3 \times 3 \times 3 \times 3 \times 3 \times 3$
12)		$=$	$= 6 \times 6 \times 6$

22) Exponents

Directions: *Write the value of each exponent.*

1) $1^3 =$

2) $1^6 =$

3) $2^2 =$

4) $2^4 =$

5) $2^5 =$

6) $2^7 =$

7) $2^8 =$

8) $2^{10} =$

9) $2^{11} =$

10) $2^{12} =$

11) $3^2 =$

12) $3^3 =$

13) $3^5 =$

14) $3^8 =$

15) $4^1 =$

16) $4^3 =$

17) $4^6 =$

18) $5^3 =$

19) $5^4 =$

20) $5^5 =$

21) $6^4 =$

22) $7^3 =$

23) $8^2 =$

24) $9^3 =$

25) $11^2 =$

26) $12^2 =$

27) $20^2 =$

28) $20^3 =$

29) $100^2 =$

30) $100^3 =$

31) $10^1 =$

32) $10^2 =$

33) $10^3 =$

34) $10^4 =$

35) $10^5 =$

36) $10^6 =$

37) $10^7 =$

38) $10^8 =$

39) $10^9 =$

40) $10^{10} =$

Critical Thinking:
Review your answers for problems 31 to 40.

41) *What pattern do you notice?*

42) *How can this pattern help you to calculate the powers of 10?*

23) Powers of 10

10^1 = **10** (1 zero)

10^2 = **100** (2 zeros)

10^3 = **1,000** (3 zeros)

10^4 = **10,000** (4 zeros)

10^5 = **100,000** (5 zeros)

10^6 = **1,000,000** (6 zeros)

10^7 = **10,000,000** (7 zeros)

10^8 = **100,000,000** (8 zeros)

10^9 = **1,000,000,000** (9 zeros)

10^{10} = **10,000,000,000** (10 zeros)

10^{15} = **1,000,000,000,000,000** (15 zeros)

10^{20} = **100,000,000,000,000,000,000** (20 zeros)

10^{25} = **10,000,000,000,000,000,000,000,000** (25 zeros)

10^{40} = **10,000,000,000,000,000,000,000,000,000,000,000,000,000** (40 zeros)

You don't need to multiply to calculate the powers of 10. The exponent correlates with the number of zeros in the value.

Example 1: What's the value of 10^4 ?

(Write a 1 with 4 zeros behind it.)

Answer: 10,000

Example 2: What's the value of 10^{12} ?

(Write a 1 with 4 zeros behind it.)

Answer: 1,000,000,000,000

The powers of 10 are easy to multiply with other numbers. When multiplying a power of 10 with a whole number:

Step 1: Write the whole number.

Step 2: Write the proper number of zeros behind it.

Examples:	Step 1:	Step 2:
2 × 10^3 =	2	2,**000**
14 × 10^5 =	14	1,4**00,000**
35 × 10^7 =	35	350,**000,000**

Directions: *Solve each problem or write the power of 10 that will make each equation true.*

1) **5 × 10^2** =

2) **9 × 10^5** =

3) **6 × 10^3** =

4) **7 × 10^{12}** =

5) **4 × 10^{14}** =

6) **12 × 10^3** =

7) **2 × 10^{10}** =

8) **30 × 10^7** =

9) **27 × 10^6** =

10) **49 × 10^9** =

11) **6 × _____ = 60,000**

12) **3 × _____ = 300,000,000**

13) **8 × _____ = 80,000,000,000**

14) **12 × _____ = 12,000**

15) **34 × _____ = 34,000**

16) **7 × _____ = 70,000**

17) **11 × _____ = 11,000,000**

18) **50 × _____ = 50,000**

19) **16 × _____ = 16,000**

20) **275 × _____ = 2,750,000**

24) Scientific Notation

Scientific notation is a system that uses the powers of 10 to make it easier to write and read long numbers. Scientific notation needs to follow two rules:

Rule 1: Scientific notation must **start with a number** that is **greater than or equal to 1** but **less than 10**.

Rule 2: That number needs to be **multiplied by a power of 10**.

Examples of Scientific Notation:	**THESE ARE NOT** written in scientific notation:
6×10^4	0.6×10^5 (0.6 is less than 1.)
2.7×10^3	27×10^2 (27 is greater than 10.)
8.045×10^5	8.045×7^5 (7^5 is not a power of 10.)

Directions: *Is it written in scientific notation? Write "Y" for yes or "N" for no beside each problem.*

1) 10×10^1 4) 9.240×10^{12} 7) 39×10^6 10) 0.9×10^3

2) 1.25×10^5 5) 11×10^7 8) 4×9^3 11) 3.8×10^7

3) 4×10^3 6) 0.8×10^2 9) 7×10^{20} 12) 5×8^2

To find the value of scientific notation, multiply the two numbers. To do this, you may move the decimal point the number of spaces to the right that corresponds to the value of the power.

6×10^4	(4 spaces right.)	6 _ _ _ _	= 60,000
2.7×10^3	(3 spaces right.)	2.7 _ _	= 2,700
8.045×10^5	(5 spaces right.)	8.0 4 5 _ _	= 804,500

Directions: *Write the value of each scientific notation in number form.*

13) 9.5×10^4 =

14) 4.703×10^3 =

15) 2×10^7 =

16) 8.25×10^5 =

17) 3.3×10^2 =

18) 5×10^{15} =

19) 1.0249×10^{20} =

20) 7.165×10^{17} =

21) 6.0017×10^8 =

22) 9×10^9 =

23) 5.28×10^7 =

24) 1.6×10^{10} =

25) 4.78831×10^6 =

25) Scientific Notation

Directions: *Write the value of each scientific notation in number form.*

1) $3.5 \times 10^4 =$

2) $7.7 \times 10^1 =$

3) $2 \times 10^9 =$

4) $8.9 \times 10^5 =$

5) $9 \times 10^7 =$

6) $1.17 \times 10^8 =$

7) $1.257 \times 10^3 =$

8) $5.28 \times 10^6 =$

9) $1.6 \times 10^{10} =$

10) $4.78831 \times 10^2 =$

Writing Scientific Notation: The first number must be 1 or greater but less than 10.
How many spaces to the left does the decimal need to move to accomplish that?
The number of spaces left corresponds to the power of 10 that is used in the scientific notation.
(Notice that **trailing zeros** are not written when writing scientific notation.)

400,000	(5 spaces left.) $4.\underline{0}\,\underline{0}\,\underline{0}\,\underline{0}\,\underline{0}$	$=$	4×10^5
62,500,000	(7 spaces left.) $6.\underline{2}\,\underline{5}\,\underline{0}\,\underline{0}\,\underline{0}\,\underline{0}\,\underline{0}$	$=$	6.25×10^7
180	(2 spaces left.) $1.\underline{8}\,\underline{0}$	$=$	1.8×10^2

Directions: *Write each number using scientific notation.*

11) 970,000,000

12) 5,600

13) 314,000

14) 73,150,000

15) 15,000

16) 605,000,000

17) 80,000,000

18) 3,210,000

19) 4,790

20) 220

21) 67,820,000

22) 5,000,000

23) 8,470,000,000

24) 43,500,000,000,000

25) 300,000,000,000,000,000

26) 150,000,000,000

27) 820,200,000

28) 71,000,000,000,000,000,000

29) 450,300,000,000

30) 900,000,000,000,000,000,000

31) 58,600,000

32) 250,000,000,000,000

33) 45,900,000,000

34) 750,000,000,000,000,000,000,000

26) Negative Exponents

$$2^{-5}$$

Base   Exponent (Power)

Repeated division can be represented with a **negative exponent**. A negative exponent is the **reciprocal** of an exponent. For example, the same way that the reciprocal of **2** is **1/2** the reciprocal of 2^5 is $1/2^5$.

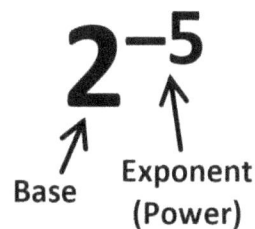

Exponential Form:	Expanded Form:	Fraction Form:	Value:
2^{-5}	$= 1 \div 2 \div 2 \div 2 \div 2 \div 2$	$= 1/2^5$	$= 0.03125$
5^{-3}	$= 1 \div 5 \div 5 \div 5$	$= 1/5^3$	$= 0.008$
20^{-2}	$= 1 \div 20 \div 20$	$= 1/20^2$	$= 0.0025$

Directions: *Fill in the blank spaces.*

Exponential Form:	Expanded Form:	Fraction Form:
Ex) 8^{-3}	$= 1 \div 8 \div 8 \div 8$	$= 1/8^3$
1) 5^{-5}	$=$	$=$
2) 12^{-4}	$=$	$=$
3) 2^{-5}	$=$	$=$
4) 5^{-3}	$=$	$=$
5)	$= 1 \div 4 \div 4 \div 4 \div 4 \div 4$	$=$
6)	$= 1 \div 15 \div 15$	$=$
7)	$= 1 \div 2 \div 2 \div 2 \div 2 \div 2 \div 2 \div 2 \div 2$	$=$
8)	$= 1 \div 6 \div 6 \div 6 \div 6 \div 6 \div 6$	$=$
9)	$=$	$= 1/7^4$
10)	$=$	$= 1/30^3$
11)	$=$	$= 1/8^6$
12)	$=$	$= 1/9^5$

27) Exponents and Calculators

Finding the value of an exponent can be time consuming. So, people often use calculators. Not all calculators have a button for calculating exponents. But, many do. If you do not have a calculator that has an exponent button on it, there are many free online calculators available.

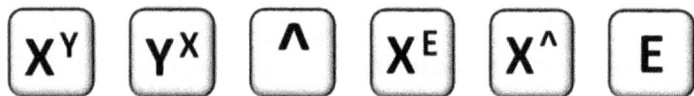

X^Y Y^X $\wedge$ X^E $X^\wedge$ E *These are examples of what an exponent button may look like.*

Directions: *Use a **calculator** to find the value of each exponent.*

1) $7^8 =$

2) $11^6 =$

3) $8^7 =$

4) $17^3 =$

5) $2^{14} =$

6) $21^4 =$

7) $3^{10} =$

8) $9^5 =$

9) $2^{11} =$

10) $6^6 =$

11) $4^{10} =$

12) $52^4 =$

13) $3^{12} =$

14) $5^6 =$

15) $2^{20} =$

16) $5^{-4} =$

17) $2^{-3} =$

18) $50^{-2} =$

19) $4^{-3} =$

20) $20^{-2} =$

21) $2^{-5} =$

22) $100^{-3} =$

23) $4^{-2} =$

24) $50^{-3} =$

25) $20^{-3} =$

26) $40^{-2} =$

27) $2^{-4} =$

28) $8^{-2} =$

29) $100^{-2} =$

30) $100^{-2} =$

---

31) $10^{-1} =$

32) $10^{-2} =$

33) $10^{-3} =$

34) $10^{-4} =$

35) $10^{-5} =$

36) $10^{-6} =$

37) $10^{-7} =$

38) $10^{-8} =$

Critical Thinking:
Review your answers for problems 31 to 38.

39) *What pattern do you notice?*

40) *How can this pattern help you to calculate the negative powers of 10?*

28) Scientific Notation with Negative Powers

Scientific notation is often used to represent large numbers, but it can represent small numbers too. Negative powers of 10 are used to represent values less than 1. <u>When multiplying a number with a negative power of 10, **move the decimal point to the left** instead of the right.</u>

2×10^{-4}	(4 spaces left.)	**0.0 0 0 2**	=	**0.0002**
5.7×10^{-5}	(5 spaces left.)	**0.0 0 0 0 5 7**	=	**0.000057**
3.145×10^{-3}	(3 spaces left.)	**0.0 0 3 1 4 5**	=	**0.003145**

Directions: *Write the value of each scientific notation in number form.*

1) $6 \times 10^{-5} =$

2) $1.5 \times 10^{-6} =$

3) $2.17 \times 10^{-4} =$

4) $8.09 \times 10^{-7} =$

5) $9 \times 10^{-1} =$

6) $6 \times 10^{-5} =$

7) $1.501 \times 10^{-9} =$

8) $8 \times 10^{-10} =$

9) $5.28 \times 10^{-3} =$

10) $1.6 \times 10^{-8} =$

11) $7.731 \times 10^{-2} =$

12) $1.501 \times 10^{-9} =$

Writing Scientific Notation: The first number must be 1 or greater but less than 10.
How many spaces to the left does the decimal need to move to accomplish that?
The number of spaces left corresponds to the power of 10 that is used in the scientific notation. <u>When the decimal is moved to the right</u>, it corresponds to a <u>negative power of 10</u>.

0.00007	(5 spaces right.)	**0.0 0 0 0 7**	=	7×10^{-5}
0.053	(2 spaces right.)	**0.0 5 3**	=	5.3×10^{-2}
0.004025	(3 spaces right.)	**0.0 0 4 0 2 5**	=	4.025×10^{-3}

Directions: *Write each number using scientific notation.*

13) **0.00006**

14) **0.0052**

15) **0.00007316**

16) **0.000000008**

17) **0.00000394**

18) **0.00015**

19) **0.056**

20) **0.00008**

21) **0.000000029**

22) **0.000000000000007**

23) **0.000000000044**

24) **0.00000932**

Name: Date: Score: *Unstoppable Owl*

29) Scientific Notation Mixed Review

Directions: *Solve each problem.*

1) $8.1 \times 10^{-5} =$

2) $7 \times 10^8 =$

3) $1.2 \times 10^2 =$

4) $1.35 \times 10^{-7} =$

5) $7.7 \times 10^3 =$

6) $3.14 \times 10^{-4} =$

7) $1.5 \times 10^{-9} =$

8) $2.2 \times 10^{10} =$

9) $2.59 \times 10^{-6} =$

10) $8.5 \times 10^{-1} =$

11) $3.8 \times 10^{-6} =$

12) $7.1 \times 10^{-3} =$

13) $1.25 \times 10^{-2} =$

14) $9 \times 10^7 =$

15) $1.1 \times 10^{-10} =$

16) $3.0247 \times 10^{-8} =$

17) $6 \times 10^{-11} =$

18) $2 \times 10^{12} =$

Directions: *Write each number using scientific notation.*

19) 350,000,000

20) 0.000572

21) 0.00000098

22) 110,000

23) 0.0077

24) 0.03

25) 0.0000121

26) 51,200,000

27) 40,000,000

28) 314,000,000,000

29) 0.000000004

30) 6,500,000,000

31) 0.000000000000066

32) 0.0000000718

33) 0.000000433

34) 850,000,000,000,000

35) 0.0000000000000019

36) 430,000,000,000,000,000

37) A virus is about 2.5×10^{-8} meters wide.
State that measurement in number form.

38) The average distance from the Earth to the Sun is about 93 million miles.
Write that distance using scientific notation.

39) Mary's fasting glucose level was 0.0000085 grams per milliliter (g/mL) of blood.
Write that measurement using scientific notation.

40) The speed of light is about 3×10^8 meters per second (m/s).
State that speed in number form.

© Libro Studio LLC

30) Square Roots

When a number is multiplied by itself, the result is a **square number**. It's called a square number because a number is often multiplied by itself when calculating the area of a square. The area for this square can be found by multiplying. ($4 \times 4 = 16$. Which is the same as $4^2 = 16$.)

Many people call 4^2 "four to the second power." But, it can also be called "four **squared**." If someone says "nine squared", it means 9^2.

4 **squared** is 16.

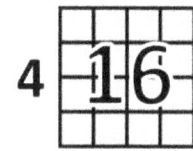

4 16
4

The **square root** of 16 is 4.

Word Form:		Exponential Form:
Four squared.	=	4^2
Thirty squared.	=	30^2
Two hundred squared.	=	200^2

Directions:
Write each problem using exponential form.

1) Eight squared. **=**

2) Forty squared. **=**

3) Six thousand squared. **=**

You've learned how to calculate the value of exponents. For example, $4^2 = 16.$ *But what if you started with 16 and wanted to find the number to the second power that equals 16?*

In this example, you'd be looking for the **square root** of 16, which is 4, because $4 \times 4 = 16$. A **square root** is the **inverse operation** of an exponent to the second power.

A **radical** can be used as a symbol for square root. Radical symbols are shown in the examples below. To use them, draw the radical symbol. Then, write the squared number inside.

Examples:

$\sqrt{9}$ = 3

$\sqrt{16}$ = 4

$\sqrt{2,500}$ = 50

Directions: *Write the value of each square root.*

4) $\sqrt{25}$ =

5) $\sqrt{81}$ =

6) $\sqrt{4}$ =

7) $\sqrt{49}$ =

8) $\sqrt{100}$ =

9) $\sqrt{64}$ =

10) $\sqrt{36}$ =

11) $\sqrt{144}$ =

12) $\sqrt{400}$ =

13) $\sqrt{10,000}$ =

14) $\sqrt{3,600}$ =

15) $\sqrt{900}$ =

31) Square Roots

Many people use calculators to help find the square root of a number. Not all calculators have a button for calculating a square root. But many do. If you do not have a calculator that has a square root button, there are many free online calculators that do.

$\boxed{\sqrt{}}$ $\boxed{\sqrt{x}}$ $\boxed{\sqrt[2]{}}$ $\boxed{\sqrt[2]{x}}$ $\boxed{\sqrt{}}$ *These are examples of what a square root button may look like.*

Directions: *Use a calculator to find each square root.*

1) $\sqrt{225}$ =

2) $\sqrt{324}$ =

3) $\sqrt{841}$ =

4) $\sqrt{784}$ =

5) $\sqrt{625}$ =

6) $\sqrt{1,369}$ =

7) $\sqrt{1,024}$ =

8) $\sqrt{1,225}$ =

9) $\sqrt{1,521}$ =

10) $\sqrt{484}$ =

11) $\sqrt{1,089}$ =

12) $\sqrt{1,156}$ =

13) $\sqrt{676}$ =

14) $\sqrt{2,500}$ =

15) $\sqrt{1,296}$ =

The square root of a **perfect square** is a <u>whole number</u>.

The square root of an **imperfect square** is <u>not a whole number</u>.

The square root of an imperfect square may be written in decimal form and are often **rounded**.

Examples:

$\sqrt{5}$ = **2.23606797749978...** = **2.24** (Rounded to the nearest hundredth.)

$\sqrt{20}$ = **4.4721359549995...** = **4.47** (Rounded to the nearest hundredth.)

Directions: *Use a calculator to find each square root. Then, round to the nearest hundredth.*

16) $\sqrt{12}$ =

17) $\sqrt{27}$ =

18) $\sqrt{48}$ =

19) $\sqrt{6}$ =

20) $\sqrt{18}$ =

21) $\sqrt{108}$ =

22) $\sqrt{80}$ =

23) $\sqrt{32}$ =

24) $\sqrt{112}$ =

25) $\sqrt{99}$ =

26) $\sqrt{75}$ =

27) $\sqrt{50}$ =

28) $\sqrt{147}$ =

29) $\sqrt{162}$ =

30) $\sqrt{500}$ =

32) Roots

Square roots are not the only roots. **Cubed roots** (**third roots**) are the inverse operation for the power of 3, **fourth roots** are the inverse operation for the power of 4, **fifth roots** are the inverse operation for the power of 5, and so on.

A radical sign and a small number on the left side of the radical are often used to symbolize these different roots. If the small number is a 4, it means it is a fourth root. If the small number is a 5, it means it is a fifth root.

(A radical sign without a small number written is assumed to be a square root. But a small 2 may be written next to the radical too.)

4 **cubed** is 64.

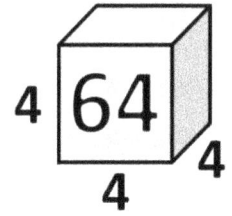

The **cubed root** of 64 is 4.

$$\sqrt[3]{64} = 4 \qquad \text{(The cubed root of 64 equals 4.)}$$

$$\sqrt[5]{243} = 3 \qquad \text{(The fifth root of 243 is 3.)}$$

$$\sqrt[7]{78{,}125} = 5 \quad \text{(The seventh root of 78,125 equals 5.)}$$

Examples of what root buttons may look like.

Some calculators will have a square root button but not a button for entering other roots. Find a calculator that does have this capability or use an online calculator that does.

Directions: *Use a calculator to find each value. When possible, <u>round to the nearest **hundredth**</u>.*

1) $\sqrt[3]{27}$ =

2) $\sqrt[4]{64}$ =

3) $\sqrt[5]{125}$ =

4) $\sqrt[3]{81}$ =

5) $\sqrt[6]{1{,}000}$ =

6) $\sqrt[3]{0.01}$ =

7) $\sqrt[5]{288}$ =

8) $\sqrt[4]{96}$ =

9) $\sqrt[3]{16}$ =

10) $\sqrt[7]{144}$ =

11) $\sqrt[6]{729}$ =

12) $\sqrt[9]{512}$ =

13) $\sqrt[7]{625}$ =

14) $\sqrt[4]{254}$ =

15) $\sqrt[5]{400}$ =

16) $\sqrt[4]{1{,}296}$ =

17) $\sqrt[7]{640}$ =

18) $\sqrt[3]{0.008}$ =

19) $\sqrt[4]{4{,}802}$ =

20) $\sqrt[6]{8{,}192}$ =

21) $\sqrt[5]{32}$ =

22) $\sqrt[4]{0.0162}$ =

23) $\sqrt[6]{7{,}776}$ =

24) $\sqrt[9]{3{,}888}$ =

25) $\sqrt[3]{216}$ =

26) $\sqrt[4]{0.0256}$ =

27) $\sqrt[9]{13{,}122}$ =

28) $\sqrt[6]{46{,}656}$ =

29) $\sqrt[5]{1{,}024}$ =

30) $\sqrt[7]{729}$ =

33) Negative Numbers with Exponents

$$(-2)^5$$

Base   Exponent (Power)

An *exponent of a negative number* and a *negative exponent* are two different things. You already learned about negative exponents. But the base number can be a negative number too. Exponents represent repeated multiplication. So, when the base number is negative, that negative number is repeatedly multiplied.

Word Form:	Exponential Form:	Expanded Form:	Value:
Negative two to the fifth power.	$= (-2)^5$	$= (-2) \times (-2) \times (-2) \times (-2) \times (-2)$	$= -32$
Negative four to the third power.	$= (-4)^3$	$= (-4) \times (-4) \times (-4)$	$= -64$
Negative nine to the second power.	$= (-9)^2$	$= (-9) \times (-9)$	$= 81$
Negative six to the fourth power.	$= (-6)^4$	$= (-6) \times (-6) \times (-6) \times (-6)$	$= 1,296$

Directions: *Fill in the blank spaces.*

	Word Form:	Exponential Form:	Expanded Form:
Ex)	Negative nine to the fifth power.	$= (-9)^5 =$	$(-9) \times (-9) \times (-9) \times (-9) \times (-9)$
1)	Negative eight to the third power.	$=$	$=$
2)	Negative ten to the fourth power.	$=$	$=$
3)	Negative twenty to the second power.	$=$	$=$
4)	Negative three to the sixth power.	$=$	$=$
5)		$= (-4)^3 =$	
6)		$= (-5)^5 =$	
7)		$= (-7)^2 =$	
8)		$= (-11)^4 =$	
9)		$=$	$= (-6) \times (-6) \times (-6) \times (-6) \times (-6) \times (-6) \times (-6)$
10)		$=$	$= (-100) \times (-100) \times (-100) \times (-100)$
11)		$=$	$= (-15) \times (-15) \times (-15) \times (-15) \times (-15) \times (-15)$
12)		$=$	$= (-3) \times (-3) \times (-3)$

34) Negative Numbers with Exponents

Directions: *Write the value of each exponent.*

1) $(-2)^2 =$

2) $(-2)^3 =$

3) $(-2)^4 =$

4) $(-2)^5 =$

5) $(-2)^6 =$

6) $(-2)^7 =$

7) $(-2)^8 =$

8) $(-2)^9 =$

9) $(-2)^{10} =$

10) $(-2)^{11} =$

11) $(-3)^2 =$

12) $(-3)^3 =$

13) $(-3)^4 =$

14) $(-3)^5 =$

15) $(-3)^6 =$

16) $(-4)^2 =$

17) $(-4)^3 =$

18) $(-4)^4 =$

19) $(-4)^5 =$

20) $(-4)^6 =$

21) $(-5)^2 =$

22) $(-5)^3 =$

23) $(-5)^4 =$

24) $(-5)^5 =$

25) $(-6)^2 =$

26) $(-6)^3 =$

27) $(-10)^2 =$

28) $(-10)^3 =$

29) $(-10)^4 =$

30) $(-10)^5 =$

Critical Thinking:

Review your answers for problems 1 to 30.

31) *What pattern did you notice that can help you determine whether an exponent with a negative base will have a negative or positive value?*

Directions: *Use the pattern to predict whether the value will be negative or positive. Write an "N" if the value will be negative or a "P" if the value will be positive.*

32) $(-5)^9$

33) $(-24)^{10}$

34) $(-8)^7$

35) $(-15)^{15}$

36) $(-9)^{18}$

37) $(-36)^{23}$

38) $(-4)^{20}$

39) $(-75)^{34}$

40) $(-10)^{46}$

41) $(-2)^{51}$

42) $(-7)^{28}$

43) $(-12)^{79}$

44) $(-50)^{31}$

45) $(-38)^{88}$

46) $(-8)^{95}$

47) $(-21)^{100}$

48) $(-6)^{170}$

49) $(-40)^{373}$

50) $(-84)^{615}$

35) Order of Operations (2-Step Problems)

Expressions often have more than one operation. The example to the right has two operations. *How would you solve it?*

It makes a difference whether you add or multiply first. But, this expression only has one correct answer, which is 14.

Example: $4 + 5 \times 2$

Should you **add first**?   Or, **multiply first**?

$$4 + 5 \times 2$$
$$9 \times 2$$
$$18$$

$$4 + 5 \times 2$$
$$4 + 10$$
$$14$$

How do you know the answers is 14? Because, people created a system of rules for solving expressions so that everyone can solve the operations in the same order and get the same answer. That's why these rules are called the **order of operations**.

The Order of Operations (PEMDAS)

Step 1: **Parentheses**	(Perform operations inside parentheses.)
Step 2: **Exponents**	(Calculate powers and roots.)
Step 3: **Multiplication & Division**	(Multiply and divide from left to right.)
Step 4: **Addition & Subtraction**	(Add and subtract from left to right.)

The order of operations requires the operations within parentheses to be performed first. Exponents and roots are calculated second. Next, multiplication and division must be performed from left to right. And finally, addition and subtraction need to be completed from left to right. If these rules are followed, everyone should calculate the same answer for the same problem.

Acronyms are often used to help remember the order of operations. **PEMDAS** and "**Please Excuse My Dear Aunt Sally**" are frequently taught to students. **BODMAS** and **BIDMAS** are often used too, because many people say "brackets" instead of "parentheses" and "orders" or "indices" instead of "exponents." Despite the differences in names, the order of operations remain the same. (Call them what you like. Just make sure to follow the rules.)

Directions: *Solve each expression using the order of operations.*

Ex) $8 + 10 \div 2$
$8 + 5$ (Divide first.)
13

1) $9 - 7 + 2$

2) $20 \div 5 \times 2$

3) $15 - 12 + 2$

4) $4 \times 7 - 3$

5) $30 - 6 \div 3$

6) $6 + 4 \times 2$

7) $12 \div 2 \times 3$

8) $22 - 2 \times 2$

36) Order of Operations (2-Step Problems)

1) $18 \div 2 + 3$

2) $6 + 4 \div 2$

3) $10 - 5 \times 2$

4) $36 \div 9 - 3$

5) $21 + 7 \times 2$

6) $40 + 10 \times 6$

7) $30 \div 6 - 1$

8) $44 - 4 \times 5$

9) $8 \times 9 - 7$

10) $100 + 50 \times 2$

11) $72 - 6 \times 6$

12) $9 + 13 \times 3$

13) $30 \div 15 + 2$

14) $49 + 7 \times 3$

15) $75 \div 15 \times 2$

16) $56 + 11 \times 8$

17) $60 + 45 \div 15$

18) $26 + 17 \times 4$

19) $245 - 10 \times 5$

20) $12 \times 8 - 6$

21) $45 \div 5 + 3$

22) $87 - 45 + 60$

23) $40 \div 5 \times 10$

24) $24 + 48 \div 12$

25) $66 \div 11 - 6$

26) $29 + 3 \times 8$

27) $90 \div 9 - 8$

28) $70 \times 10 - 7$

29) $90 \div 5 \times 4$

30) $615 + 300 \div 30$

37) Order of Operations (3-Step Problems)

It's time to practice 3- step problems. The same rules for the order of operations apply. Think about each problem carefully and decide which operations to perform first.

Directions: *Solve each expression using the order of operations.*

1) $12 \div 2 - 2 \times 0$

2) $26 + 5 - 12 \times 2$

3) $9 \times 8 - 15 \div 5$

4) $80 \div 8 + 2 \times 11$

5) $49 - 14 + 7 \times 5$

6) $78 + 54 \div 9 \times 7$

7) $56 \div 8 - 36 \div 6$

8) $5 \times 6 - 18 + 9$

9) $81 \div 9 + 20 \times 5$

10) $9 \times 5 - 15 + 47$

11) $52 + 63 \div 7 \times 9$

12) $168 + 60 \div 12 - 10$

13) $235 - 14 + 3 \times 5$

14) $200 \div 10 - 20 + 65$

15) $235 - 30 + 40 \times 5$

16) $53 + 80 - 4 \times 3$

17) $114 + 30 - 36 \div 12$

18) $72 \div 8 + 11 \times 4$

19) $9 \times 6 - 20 + 25$

20) $99 + 45 \div 9 + 15$

21) $72 \div 8 - 4 \times 2$

38) Order of Operations (3-Step Problems)

1) $2 \times 7 - 45 \div 5$

2) $19 + 72 \div 6 + 15$

3) $30 \times 8 - 130 + 60$

4) $48 \div 6 + 7 \times 4$

5) $110 + 43 - 72 \div 8$

6) $36 \div 4 - 4 \times 2$

7) $69 + 127 - 8 \times 7$

8) $413 - 200 + 5 \times 10$

9) $9 \times 2 - 72 \div 9$

10) $28 + 74 - 5 \times 2$

11) $61 + 35 - 72 \div 8$

12) $5 \times 10 - 6 \div 2$

13) $12 + 8 \div 4 - 1$

14) $100 \div 10 + 50 \times 2$

15) $80 - 40 \div 4 \div 2$

16) $490 - 140 - 12 \times 12$

17) $25 \div 5 \times 9 - 15$

18) $10 \times 6 - 15 \div 5$

19) $10 + 100 \div 100 - 9$

20) $201 - 190 + 2 \times 0$

21) $99 \div 9 - 3 \times 3$

22) $11 \times 8 + 24 \div 2$

23) $24 - 12 \div 3 \times 4$

24) $98 + 72 \div 6 - 10$

39) Order of Operations (3-Step Problems)

1) $12 \times 4 - 54 \div 9$

2) $11 \times 7 + 34 - 10$

3) $69 + 51 - 8 \times 4$

4) $3 \times 50 - 30 + 20$

5) $16 + 100 \div 10 - 14$

6) $35 - 63 \div 9 \times 4$

7) $7 \times 2 + 88 \div 11$

8) $64 \div 8 - 8 \times 1$

9) $50 \div 2 \times 5 - 8$

10) $132 \div 12 - 6 \times 0$

11) $14 + 36 \div 2 - 9$

12) $100 - 97 + 8 \times 10$

13) $12 \times 3 - 32 \div 4$

14) $42 \div 6 - 3 \times 1$

15) $7 \times 3 + 9 - 11$

16) $24 + 14 - 9 \times 2$

17) $20 \times 4 - 27 + 8$

18) $6 + 144 \div 12 - 12$

19) $96 - 60 \div 6 \times 4$

20) $8 \times 11 + 48 - 12$

21) $7 \times 8 + 4 \div 4$

22) $63 \div 7 - 3 \times 3$

23) $19 + 14 \div 14 - 18$

24) $51 - 19 + 8 \times 4$

40) Order of Operations (3-Step Problems)

1) $6 \times 3 - 40 \div 20$

2) $21 + 49 \div 7 + 32$

3) $9 \times 8 - 23 + 11$

4) $45 \div 9 + 12 \times 4$

5) $64 + 16 - 80 \div 10$

6) $48 \div 4 - 6 \times 2$

7) $157 + 121 - 40 \times 6$

8) $96 - 80 + 9 \times 5$

9) $35 \div 5 - 2 \times 0$

10) $55 \div 5 - 2 \times 4$

11) $98 + 28 \div 7 \times 9$

12) $4 \times 9 - 35 \div 5$

13) $70 \div 10 - 8 + 19$

14) $99 \div 9 + 11 \times 3$

15) $48 \div 4 - 24 \div 4$

16) $49 - 14 - 6 \times 5$

17) $70 \div 7 \times 2 - 20$

18) $121 \div 11 - 2 \times 4$

19) $16 + 88 \div 8 - 9$

20) $32 - 17 + 4 \times 6$

21) $84 \div 7 - 6 \times 2$

22) $1 \times 13 + 25 \div 5$

23) $95 - 15 \div 5 \times 3$

24) $33 + 72 \div 6 - 10$

41) Order of Operations (Exponents)

Exponents must be calculated before multiplication or division takes place. Review the chart on page 35 if you want a refresher.

Example:
$$8 - 4 \times 3^2 \div 6$$
$$8 - 4 \times 9 \div 6$$
$$8 - 36 \div 6$$
$$8 - 6$$
$$\mathbf{2}$$

Directions: *Solve each expression using the order of operations.*

1) $38 - 2 \times 4^2 \div 4$

5) $3^2 + 5^2 \times 4 \div 2$

9) $3^2 + 4^2 \div 8 \times 2$

2) $3^2 \times 5 - 2^2 + 6$

6) $6^2 - 5^2 \times 2 \div 10$

10) $5^2 \times 1 + 3^2 - 11$

3) $9^2 + 9 \times 2 - 18$

7) $13 - 2^2 + 1 \times 2^3$

11) $3^2 \times 2^3 - 55 \div 5$

4) $2^6 + 2^3 - 8 \div 8$

8) $4^2 \div 2 - 0 \times 2^4$

12) $19 - 4^2 \div 8 \times 3$

Unstoppable Owl

42) Order of Operations (Exponents)

1) $3^2 + 5^2 - 5 \times 4$

2) $2^3 \div 2 + 17 \times 1$

3) $9^2 \times 0 + 2^4 - 4$

4) $11 + 3^3 \div 3^2 + 7$

5) $8^2 \div 2^3 - 4 \times 2$

6) $10^2 \div 4 \times 2 - 2^2$

7) $8^2 \div 2 - 1 \times 2^2$

8) $2^4 + 5^2 \div 5 - 10$

9) $2^5 - 17 + 2^2 \times 6$

10) $3^4 \div 9 - 3^2 \times 1$

11) $3^3 + 12 - 10^2 \div 10$

12) $64 \div 2^3 + 7 \times 2^2$

13) $9 \times 6 - 29 + 2^5$

14) $4^2 + 81 \div 3^2 + 15$

15) $88 \div 2^3 - 3 \times 2$

43) Order of Operations (Exponents)

1) $6^2 + 0 - 2^3 \times 3$

2) $3^3 + 7 - 1^5 \div 1$

3) $7^2 \times 1 - 4^2 + 3$

4) $15 + 8^2 \div 8 + 11$

5) $9^2 \div 9 - 2^2 \times 2$

6) $6^2 \div 6 + 14 \times 1$

7) $9 \div 3^2 + 6 \times 2^2$

8) $7^2 \times 1^{10} - 2^2 + 5$

9) $4^2 + 90 \div 3^2 + 14$

10) $80 \div 2^3 - 3 \times 2$

11) $7^2 \div 7 \times 4 - 3^3$

12) $2^4 \div 8 - 0 \times 2^2$

13) $5^2 + 5^2 \div 5 - 13$

14) $8^2 - 20 + 2^2 \times 10$

15) $4^2 \div 2^3 + 2^2 \times 1$

44) Order of Operations (Exponents)

1) $39 - 9 \div 1^7 \times 2^2$

2) $3^3 \times 8 - 2^4 \div 2$

3) $4^2 \div 8 - 2^2 + 39$

4) $2^4 \times 1^9 \div 8 + 57$

5) $9^2 + 9 - 56 \div 2^3$

6) $72 \div 8 - 4 \times 1^3$

7) $5^2 \div 5 - 1 \times 2^2$

8) $9^2 + 7^2 \div 7 - 12$

9) $2^2 \times 2^3 + 8^2 - 60$

10) $24 \div 2^3 + 2^2 - 1$

11) $6^2 \div 9 + 9 \times 1$

12) $9 - 3^2 + 8 \times 2^2$

13) $7^2 + 1^{10} - 2^2 \times 5$

14) $9^2 + 40 \div 2^2 + 9$

15) $25 \div 5^2 - 1^9 \times 1$

45) Order of Operations (Exponents)

1) $55 + 6^2 - 3^2 \times 3$

2) $5^2 + 4^2 - 2^2 \div 1$

3) $7^2 - 40 - 6^2 \div 4$

4) $64 + 10^2 \div 10 - 31$

5) $54 \div 3^2 - 1^8 \times 3$

6) $63 \div 7 \times 5 - 3^3$

7) $0 \div 12 - 8 \times 2^2$

8) $100 + 7^2 \div 7 - 3^3$

9) $65 - 8^2 + 2^2 \times 5$

10) $4^2 + 2^3 - 6 \times 2^2$

11) $9^2 - 8 \times 5 + 1$

12) $90 \div 3^2 + 6 \times 1^{12}$

13) $2^4 + 8 - 24 \div 2^2$

14) $8^2 - 49 \div 7^2 + 16$

15) $35 \div 7 - 2^2 \times 1^9$

Unstoppable Owl

46) Order of Operations (Parentheses)

With the order of operations, **parentheses** are a priority. Operations inside the parentheses must be done first. The order of operations applies inside of parentheses too. (The exponent is calculated before multiplying, which is done before subtraction.)

Example:
$(50 - 4 \times 3^2) \div 2$
$(50 - 4 \times 9) \div 2$
$(50 - 36) \div 2$
$14 \div 2$
7

Directions: *Solve each expression using the order of operations.*

1) $12 - (3 \times 2^2) \div 4$

2) $(3^2)(25 - 2^2 \times 6)$

3) $(9^2 \div 3^2 + 2) \times 7$

4) $(64 \div 2^3) \times 9 - 6^2$

5) $66 \div 6 \times (5^2 - 2^4)$

6) $(36 - 5^2 + 3) \div 7$

7) $(4 + 5^2 \div 5)^2 + 19$

8) $(2^3 \times 6 - 3) \div 3^2$

9) $48 \div (2^2 + 2)(7)$

10) $44 + (37 - 28) \times 6$

11) $91 + (3^2 \times 10 + 11)$

12) $(6^2 - 5^2)(8 + 3)$

47) Order of Operations (Parentheses)

1) $50 + (3^2 \times 7) - 4$

2) $(3^2 - 5)(2^2 \times 3)$

3) $(45 \div 3^2 - 2)\, 5$

4) $28 + (2^3 \times 7) - 6$

5) $(11 - 15 \div 5)^2 + 19$

6) $(42 \div 7 + 2)(5 + 1^5)$

7) $44 - (201 - 196)\, 7$

8) $75 + (2^2 \times 11 - 14)$

9) $(82 - 71)(6 + 3)$

10) $(8)(36 \div 9 - 2^2)$

11) $(30 - 5^2)(21 \div 7)$

12) $(63 \div 7 - 5)^2 + 86$

13) $(8 \times 6 - 21) \div 3^2$

14) $43 + 4\,(83 - 78)$

15) $(27 \div 9 + 2)(3^2 + 1^6)$

48) Order of Operations (Parentheses)

Multiplication can be symbolized in many ways. A • symbol is often used instead of an × to signify multiplication. *Why use a dot*? Because, letters are used in algebra, and the multiplication symbol (×) is often confused with the letter X. Using a dot helps to avoid confusion.

Example:
$(30 + 6 \bullet 3) \div 4$
$(30 + 18) \div 4$
$(48) \div 4$
12

Directions: *Solve each expression using the order of operations.*

1) $(15 + 6 \bullet 8) \div 7$

2) $(3^2 \bullet 11 - 74) + 25$

3) $(4^2 - 6)(5 \bullet 2)$

4) $77 + (85 - 79) \bullet 6$

5) $9^2 + (3^2 \bullet 16 \div 8)$

6) $134 - (3^2 \bullet 18 \div 9)$

7) $(65 - 8^2 + 1) \bullet 2^3$

8) $(7^2 \bullet 1^{10} - 3^2) - 40$

9) $6^2 + (15 \div 5) \bullet 7$

10) $(10^2 \bullet 1 + 3^2) - 54$

11) $3^2 \bullet (2^3 - 66 \div 11)$

12) $(22 - 4^2) \div (2 \bullet 3)$

49) Order of Operations (Parentheses)

Multiplication can also be symbolized without using a symbol. *How is that possible?* Whenever you don't see a symbol between two numbers, it means you must multiply them. (Typically, there will be parentheses around one of the numbers to make it distinct.)

Example:

$(9 - 8 \div 2)\,6$

$(9 - 4)\,6$

$(5)\,6$

30

Directions: *Solve each expression using the order of operations.*

1) $(55 - 2 \bullet 5^2)\,4$

5) $(1^5 + 2^2)\,(24 \div 3)$

9) $(2^2 + 4^2 \div 16)\,7$

2) $3^2\,(11 - 3^2 + 4)$

6) $250 - (5^2 \bullet 2 + 100)$

10) $9\,(1 + 8^2 - 54)$

3) $(6^2 + 4 \bullet 2) - 28$

7) $(6 + 3^2 - 7) \bullet 2^3$

11) $(2^3 \bullet 3^2) - (90 \div 10)$

4) $2^5\,(2^3 - 8) \div 16$

8) $50\,(2 - 0 \bullet 5^3)$

12) $158 - (7^2 \div 49)\,8$

50) Order of Operations (Parentheses)

1) $(5^2 + 10 - 3^2)\ 2$

2) $(7^2 + 31) - (1^5 \bullet 24)$

3) $6\ (25 - 4^2 - 5)$

4) $(70 - 8^2)\ (32 \div 4)$

5) $44 \div (47 - 6^2) \bullet 2$

6) $9^2 + (8 \bullet 4 - 26)$

7) $(4^2 \div 2 - 1) \bullet 2^2$

8) $82 + (60 \div 5) - 33$

9) $(5^2 - 20 + 2^2)\ 10$

10) $(3^2 \bullet 11) - (57 \times 1)$

11) $54 + (6 \bullet 5)\ 3$

12) $9\ (34 - 7 \bullet 2^2)$

13) $(9^2 \bullet 1^{10} - 7^2) + 105$

14) $4^2 + (20 \div 2^2 + 14)$

15) $(40 \div 2^2) + (9 \times 7)$

51) Order of Operations (Nested Parentheses)

There can be more than one set of parentheses. There can even be sets of parentheses inside of parentheses, which are known as **nested parentheses**. The operations within nested parentheses must be completed first. Then, the operations within the outer set of parentheses should be completed next.

Example:
$$3 + ((14 + 6) \div 4)\,2$$
$$3 + (20 \div 4)\,2$$
$$3 + (5)\,2$$
$$3 + 10$$
$$13$$

Directions: *Solve each expression using the order of operations.*

1) $8 - (2 \bullet (16 \div 4))$

2) $(88 - (10 - 2)) \div 8$

3) $(4 + (10 - (3 + 5)))\,7$

4) $70 + (2\,(9 \div 3 + 2)^2)$

5) $183 + (3\,(7 - 4)^2)$

6) $5\,((3^2 \div 3)^2 + 6)$

7) $87 - ((24 \div 8)^2 + 31)$

8) $(9 + (16 + 10))\,5$

9) $((1^{10} + 3^2) \div 5)\,7$

10) $((12 - 3^2)\,2)^2 - 31$

11) $3\,((8 + 2) \bullet 5 - 45)$

12) $18 - (1 + (3^3 - 22))$

52) Order of Operations (Nested Parentheses)

1) $(92 - ((14 - 8) \div 2)\,3)$

2) $(4\,(3 + 7) - 6^2) \div 2$

3) $((7^2 \bullet 1) - 35) + 8\,(6)$

4) $(34 - (10^2 \div 100)) + 9$

5) $((8 \bullet 9) \div 8)^2 - 31$

6) $111 + (30\,(36 \div 9))$

7) $(56 \div (2^2 + 3)) \bullet 4$

8) $(2\,(103 - 96)) + 46$

9) $(1^{10} + (40 \div 5)) \bullet 6$

10) $20 \div (2^3 - (2^2 \bullet 1))$

11) $(35 \div (21 - 14))\,6$

12) $((16 \div 2^4) + 4) - 5$

13) $13 \bullet (21 - (4^2 + 5))$

14) $16 + (8\,(25 \div 5^2) + 9)$

15) $(80 \div (57 - 49)) \bullet 6$

53) Order of Operations (Nested Parentheses)

1) $((54 - 29) - 2^2) \div 7$

2) $(5 + (16 - (4 + 3^2))) \, 7$

3) $45 \div ((11 - 2^3) \, 3)$

4) $105 - (4 \, (19 - 11))$

5) $26 + ((15 - 9) \, 2 - 12)$

6) $(49 - 7 \, (31 - 5^2)) \bullet 9$

7) $(35 \div (8 - 1)) \bullet 2^2$

8) $((25 + 5^2) \div 10)^2 - 19$

9) $(8(6) - (12 + 6^2)) \bullet 13$

10) $((28 \div 4) \, (36 - 9)) + 2$

11) $(40 \div (7 + 13)) \, 9$

12) $49 \div (3^2 + (10 \times 2^2))$

13) $(8 \, (4^2 - 2^2)) + 3(7)$

14) $((5 + 4) \, (16 - 9)) \div 7$

15) $6 \, (2 \, (19 - 2^3) - (4 \bullet 3))$

54) Order of Operations (Nested Parentheses)

1) $(3^2 \cdot (28 \div 7)) - 5(4)$

2) $(9^2 (10 - 9) - 54) + 67$

3) $112 + (4^2 - 2(25 \div 5))$

4) $((30 - 16) \div 2) \cdot 3$

5) $100 \div ((29 - 24) \, 2)$

6) $59 - ((63 - 6^2) + 15)$

7) $(7 \, (32 \div 2^3) - 25) \cdot 2^2$

8) $165 + (9(21 \div 7) - 17)$

9) $(22 - (13 + 2^2)) \, 9$

10) $(7(16 \div 2^3) - 2^2) \, 5$

11) $(5 \, (42 \div 6)) - 8(4 + 2)$

12) $32 \div ((48 \div 6) \, 2^2)$

13) $((68 \times 1^{10}) - 47) + 23$

14) $((29 + 11) \div 2^2) + 89$

15) $((54 \div 3^2) - 3) \, 6$

55) Order of Operations (Nested Parentheses)

1) $((2^3 \times 6) \div (3 + 5))\ 9$

2) $(9 + (3 + 6)^2) - 37$

3) $314 - (3\ (2 \times 3) + 25)$

4) $(2 + (10^2 - 94)^2) + 40$

5) $26 + ((64 \div 8) - 4)^2$

6) $(2 + (54 \div 3^2)) \cdot 7$

7) $6^2 + ((1 + 7^2) - 11)$

8) $(2^3 + (92 - 60)) \div 5$

9) $((25 - 4^2)\ 2) + 3(6 + 4)$

10) $102 - (47 - (36 \div 6))$

11) $((30 \cdot 2) - 27) + 2^3$

12) $(6 + (66 \div 11)) - 12$

13) $((108 - 60) \div 6) \div 2^2$

14) $(8(16 - 7) + 48) - 12$

15) $7(2^3) + (5(2)\ (28 \div 4))$

56) Order of Operations (Nested Parentheses)

1) $19 + ((8 - 2^3) \times 3)$

2) $3^3 + (35 \div (1^5 \times 7))$

3) $((164 - 85) - 4^2) + 39$

4) $(95 + (20 \div 4)) - 3$

5) $79 - ((24 \div 4) \, 4 + 12)$

6) $((6(3) + 45) \div 63) + 1$

7) $(((12 \div 4) \, 9) \div 3) + 12$

8) $(7(42 \div 6) - 3) + 1^{10}$

9) $7(((3 + 10) - 11) \div 2)$

10) $(31 + (18 - 4^2) \, 5) - 22$

11) $9(7) \div (25 - (2 \times 9))$

12) $((34 - 5^2)^2 + 40) - 99$

13) $((21 \div 3)6 + 26) - 54$

14) $((26 + 34) \div 6) + 54$

15) $((56 \div 2^3) - 4) \, 2 - 5$

57) Order of Operations (Nested Parentheses)

1) $(8 \, (11 - 3^2)) + 5 \, (4 \bullet 9)$

2) $((122 - 67) + 26) \div 9$

3) $(10 + (24 \div 8) - 3) \, 6$

4) $(10 \, (4 + 2) - 12) \div 2^3$

5) $(51 + 4 \, (23 - 4^2)) - 62$

6) $(19 + (1 + 7)^2) - 40$

7) $500 - (5(4 \times 2) + 125)$

8) $(11 + (6^2 - 31)^2) - 35$

9) $26 + ((42 \div 6) - 5)^2$

10) $88 - (4(55 \div 11) + 22)$

11) $(5 + (8 \div 2^2)) \bullet 9$

12) $((46 + 4) - 7^2) - 1^8 + 3$

13) $3^3 + ((92 - 52) \div 4)$

14) $(2 \, (35 - 26) + 3) - 2^3$

15) $112 - (16 - (14 \div 7))$

58) Order of Operations (Negative Numbers)

It's time to use the order of operations with expressions that contain negative numbers. Remember, parentheses are often placed around negative numbers to help distinguish them from the other operations around them.

Example:
$$(-8 - 2) \times (-4)^2 \div 2$$
$$-10 \times (-4)^2 \div 2$$
$$-10 \times 16 \div 2$$
$$-160 \div 2$$
$$-80$$

Directions: *Solve each expression using the order of operations.*

1) $-5 + -3 \times -6 \div 2$

2) $(-3)^2 \times 5 - (-2)^2 + 6$

3) $2^2 + -9 \times 2 - (-10)$

4) $-4 + (-2)^5 - 12 \div -3$

5) $(-2)^2 + (-5)^2 - 18 \div -2$

6) $6^2 + 3^2 \times -2 \div 9$

7) $-13 + 2^2 + 1 \times (-2)^3$

8) $-4^2 \div 8 - (-5) \times 3$

9) $-(4^2 + 3^2) \div 5 + -7$

10) $12 - (-9) - 3^2 + 11$

11) $9 + (8 \div -2 - 3)(-5)$

12) $7(24 \div -6 \times 2) - (-56)$

59) Order of Operations (Negative Numbers)

1) $(7^2 - 45 + 2^2)(-6)$

2) $(-8)^2 + 55 + 1^5 \cdot -25$

3) $-6(25 - 4^2 - 4) + -22$

4) $-(70 - 8^2)(-36 \div 4)$

5) $-45 \div (41 - 6^2) \cdot (-2)$

6) $6^2 - 7 \cdot 3 + -26$

7) $((-2)^3 \div 2 - 6) \cdot 2$

8) $58 + -(12 \div 2) - 33$

9) $3^2 - 29 + 2^3 - (-17)$

10) $(-3)^2 \cdot 11 - 77 \cdot -1$

11) $24 - (7 - 9)(-3)$

12) $(9 - 34 + 17) \cdot 2^2$

13) $(-2)^3 - 1^6 + 7^2 - 15$

14) $4^2 - (-28 \div 7 + 14)$

15) $(35 \div -7) - (2 \times 9)$

60) Order of Operations (Negative Numbers)

1) $-81 \div -9 - (-2) \bullet 3$

2) $29 - 63 \div 7 - (-15)$

3) $3 \bullet -8 - 13 + 37 - 11$

4) $48 \div 8 + -7 \bullet 4 + -6$

5) $105 + -66 - 72 \div 9$

6) $-24 - 12 \div 3 \times -4$

7) $112 + (-120) - 7 \bullet -8$

8) $-10 \div 10 - (-3) \times 11$

9) $-57 + 42 \div 6 \bullet 7$

10) $(-4)^2 + 29 - 38 \bullet 1$

11) $-(-5)^2 \div -5 \bullet 3^2 - 15$

12) $-69 + 120 - 2^3 \bullet 7$

13) $(-2)^3 \bullet (-1)^5 + 6^2 - (-30)$

14) $54 \div 6 - 4 \bullet -5$

15) $10^2 - (-28) \div 7 - 104$

61) Order of Operations (Negative Numbers)

1) $41 + 56 \div -7 + (-32)$

2) $-80 \div 8 \bullet -2 - 20$

3) $-(-4)^2 + 88 \div 8 + (-9)$

4) $-9(-1) + -6 + 45 \div -5$

5) $17 + -16 - 50 \div -5$

6) $-79 + 21 \div -7 \bullet -5$

7) $250 - 125 + -60 \bullet 3$

8) $-32 \div 4 - 2 \times 0 + 4$

9) $-39 - 11 + -4 \bullet 6$

10) $-12 - 1 \bullet 9 + (-4)^2 + 37$

11) $-(-6)^2 - 5 + 12 - 4$

12) $-33 + 85 - 9 \bullet (-2)^3$

13) $(-2)^4 - 22 - (-7) \bullet 2$

14) $27 \div -3 - (-5) - 96$

15) $-31 - (-26) - (-8) \div 4$

62) Order of Operations (Negative Numbers)

1) $-16 \div -2 + (-4) \cdot 9$

2) $-127 + 78 - 6 \div 2 - 8$

3) $77 - 113 - (-24) - 13$

4) $4^2 - 17 - 36 \div (-2)^2$

5) $15 + -30 - (-18) \div -9$

6) $(-9 - 5 + 17 - 2)^2 (-6)$

7) $314 + (-520) - 9 \cdot 2^3$

8) $- (-9) \times 11 - 10^2 \div 10$

9) $-9 (-1 + 8^2 - 54) - 7$

10) $-14 - 3(2 - 3) + 27$

11) $-3 \cdot 2^3 + 20 - 82$

12) $-41 - 3^3 - (-7) + (-6)$

13) $-3^3 + (-1)^5 + 7^2 - 40$

14) $(5^2 - 23 + 2)^2 - (-13)$

15) $75 - (-9)^2 + 7 - (-23)$

63) Order of Operations (Challenge Problems)

New skills are not introduced in this section. The expressions are just longer and will require more steps to complete. They're designed to be challenging. Take your time with them. The better you understand the order of operations, the easier it will be to learn algebra and other mathematical skills in the future.

Directions: *Solve each expression using the order of operations.*

1) $15 + 2\,(6 + ((40 \div 8) - 3)\,4 - 2)$

2) $(42 - ((16 - 2^3) \div -2)\,3) + (-2) - 34$

3) $((4\,(-3 + 5) - 4^2) \div 2) + 7\,(-4 + 5)$

4) $6((27 - (2^4 + -5)) - 10) + - (2 \bullet -5)$

5) $-8(16 + (8\,(-25 \div 5^2) + 9)\,(-7))$

6) $(-48 \div (55 - 49))\,(5) + 8 - (-33)$

64) Order of Operations (Challenge Problems)

1) $((4 + (-63 \div 3^2)) \bullet -7) - (-37 + 20)$

5) $((-19 + (4 + 3)^2) - 40)^2 \, (-1 + (2^3))$

2) $-46 + ((-1 + 3^2)^2 - 59) + (-3 \times 7)$

6) $(175 - (5(-5 \bullet 2) + 125)) + -1000$

3) $-115 - (104 - 69) \div -7 + (-246)$

7) $(-166 + (722 - 547))^2 - 46 + (-37)$

4) $(((36 - 6^2) - 5) - 3)^2 - (-12 + 4)^2$

8) $(270 + ((48 \div -6) + 4)^2 - 261) \div -5$

65) Order of Operations (Challenge Problems)

1) $-9(2^3) \div (-4(78 - (10 \times 8))) + -121$

2) $(((84 - 100) + 4 \bullet 10) - 99) + -600$

3) $(((21 \div -7)6 + 26)^2 - 54)^2 + -99$

4) $((-17 + 35) \div -6)(40 \div -8) - 7 \bullet 6$

5) $(-19 + ((16 - 4^2) \times 3) + 15)^2 + -94$

6) $(-3^3 + (-35 \div (-1^5 \times 5))^2 - 20)^2 - 4$

7) $((264 - 185) + 16) - 24 - 2(56 \div 7)$

8) $(-9(-1 + 48 - 39) + 55) + 3^2 \div -9$

66) Order of Operations (Challenge Problems)

1) $(((-34 - 2) \div 2^2) + -19) - 3(2 - 4(3))$

5) $((6(-3) + 68) \div 10) - 5(-1 - 2) \div 3$

2) $((54 \div -6 - (-7)) \cdot 6) - (24 \div 3)$

6) $((((-54 + -46) \div 10)4) \div 5) - 12 - 8$

3) $(((36 + 34) \div -10)^2 + -54)^2 + 3(-9)$

7) $(9(-48 \div 8) - 3) - 1^{10} - 2(9 - 15)$

4) $-3(19 - (-5^2)) - (7((-32 \div 2^3) - 4))$

8) $9((((-13 - 15) \div 4) + 17) \div -2) - 5$

67) Order of Operations (Challenge Problems)

1) $((99 - 57) \div -6 + 13) - ((-2)^3 + (-1)^5)$

5) $(((6(-2) \div 2^2)(-8)) \div 4) - 6 + -19$

2) $-43 - (((106 + 95) - 187) \div 2)(-6)$

6) $((18 + 16 - 5^2)^2 - 82) + (9 + -108)$

3) $(((8 \cdot -3) \div 2 - 2^4 \cdot 0)3 + -10) + 11$

7) $2^4(((-6 - 7 - 2^2 - 1 \times 2^3) \div 5) + 4)$

4) $(2(-303 + 292) - 42) \div (-2 \cdot 3 + -2)$

8) $(203 - 128) + (2^2(8 + 3) - 24) - 56$

Name: Date: Score: *Unstoppable Owl*

68) Order of Operations (Challenge Problems)

1) $-7(69 - (10 \times 8)) + 2(-505 + 497)$

5) $(((177 - 194) + 3 \bullet 10) - 101) + -2$

2) $-6 + -5(((83 - 9^2) \times 3 + -4)^2 - 12)$

6) $-27 + ((-15 \div (1^6 \times 5))^2 - 3 \bullet 4)^2 - 4$

3) $((((91 - 63) \div -7)6 + 26)^2 - 3)^2$

7) $(2(-74 + 66) + 48 - 115) - (3^2 \bullet -9)$

4) $44 \div -4 - 7 - 5((-89 + 71) \div -6)$

8) $(9(-1 + 114 - 105) - 53 - 3^2) \div -5$

69) Order of Operations (Challenge Problems)

1) $-29 + ((78 - 24) \div 3^2)\,(-2) + -60$

5) $8((6^2 \div 9)^2 - 16)2^4 + -2(46 - 53)$

2) $(-5((2^2 \bullet (-3)^2) \div 6) + -11) - 5^2 + -1$

6) $2^3(500 - 603 + 96) \div 7 - 4 \bullet -2$

3) $(3^2 - (21 - 43) - 15) - 5^2 - 4 \div -2$

7) $((302 - 289) - 4^2)^2 + -1 \times 2^3 - 37$

4) $(((-6 + -2) + (-92 + 60)) \div 8)\,(-4)$

8) $(-5(993 - 1000) - 89) \div 6 + -3(6)$

70) Order of Operations (Challenge Problems)

1) $(((-47 + 23) \div 6)^2 + -72 \div 3^2)^2$

5) $((-19 + (4 + 3)^2) - 40)^2 (-1 + (2^3))$

2) $-79 + ((2 + -5)^2 - 69) - (103 + 36)$

6) $(275 - (3(4 \cdot -2) - 121)) - 500$

3) $(((-121 + 11^2) - 2) - 4)^2 - (12 - 4)^2$

7) $(171 + (725 - 900))^2 + 96 + (-10)^2$

4) $-117 - 4((-104 + 59) \div -5) + -28$

8) $(301 + ((-32 \div 4) + 1)^2 - 30) \div -8$

71) Ratios

Ratios and fractions have a lot of similarities. They both compare one thing to another thing. Ratios can even be written in fraction form. But ratios and fractions are not the same thing.

Fractions represent parts of a whole.
- The things being compared must be the same.
- Fractions can be added.

Example:

¾   +   ½   =   1 ¼

Three-fourths of a cookie plus one-half of a cookie equals 1 ¼ of a cookie.

Ratios may compare two different things.

Example:

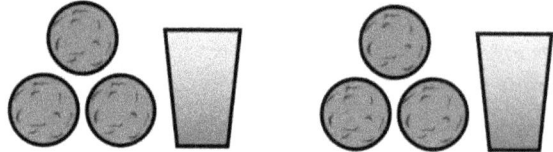

The ratio of cookies to cups is **3 to 1**.
The ratio of cups to cookies is **1 to 3**.

You can't add cookies and cups.
They are two different things.
The answer would not make sense.

Ratios can be written in a few different ways. They can be written in word form, written with a colon, or written as a fraction.

Word Form:	With a Colon:	As a Fraction:
3 to 1	3:1	3/1
4 to 7	4:7	4/7
12 to 5	12:5	12/5

Directions: *Write each ratio:*

With a colon.

1) **5 to 8 =**

2) **2 to 7 =**

3) **6 to 1 =**

4) **7 to 13 =**

5) **15 to 11 =**

6) **20 to 3 =**

7) **2 to 5 =**

8) **14 to 30 =**

9) **100 to 9 =**

10) **25 to 80 =**

In word form.

11) **6:7 =**

12) **13:5 =**

13) **4:10 =**

14) **1:8 =**

15) **60:100 =**

16) **13:30 =**

17) **51:19 =**

18) **18:25 =**

19) **7:5 =**

20) **24:12 =**

In fraction form.

21) **8:3 =**

22) **9:16 =**

23) **25:75 =**

24) **16:32 =**

25) **1:8 =**

26) **81:27 =**

27) **6:19 =**

28) **15:3 =**

29) **7:18 =**

30) **5:100 =**

72) Ratios

Directions: *Write each ratio shown in <u>word form</u>.*

1) What is the ratio of
   squares to circles?

2) What is the ratio of
   circles to squares?

3) What is the ratio of
   ovals to triangles?

4) What is the ratio of
   triangles to ovals?

A ratio can be **reduced** (also called **simplified**) the same way a fraction can be reduced. If you already know how to reduce fractions, learning to reduce ratios should be easy.

Reducing fractions was taught in the workbook shown to the right. The examples on this page show how to reduce ratios. If you haven't already completed this workbook, you may want to for extra practice.

To reduce a ratio to its **simplest form**, <u>divide both numbers in the ratio by the same divisor, until the ratio cannot be reduced any further</u>.

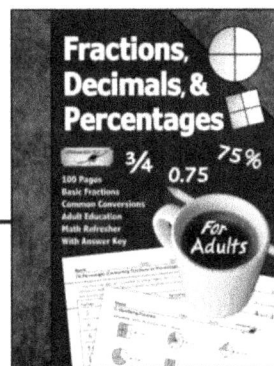

Write each ratio in simplest form using colons.

Example 1: $\dfrac{8 \text{ apples}}{14 \text{ lemons}} = \dfrac{8 \text{ apples} \div 2}{14 \text{ lemons} \div 2} = \dfrac{4 \text{ apples}}{7 \text{ lemons}} = 4{:}7$

Example 2: $\dfrac{30 \text{ cars}}{75 \text{ bikes}} = \dfrac{30 \text{ cars} \div 3}{75 \text{ bikes} \div 3} = \dfrac{10 \text{ cars}}{25 \text{ bikes}} = \dfrac{10 \text{ cars} \div 5}{25 \text{ bikes} \div 5} = \dfrac{2 \text{ cars}}{5 \text{ bikes}} = 2{:}5$

In example 1, both numbers can be divided by 2 and only one round of division was used to reduce the ratio to its simplest form.

In example 2, both numbers could be divided by 5. Then, both numbers could still be divided by 3. Two rounds of division were used to reduce this ratio. The ratio could also have been reduced by dividing both numbers by 15. Then, only one round of division would have been needed. Either way, the answer is the same. 30:75 reduced to its simplest form is 2:5.

Directions: *Write each ratio in simplest form using colons.*

5) **36:4** =

6) **2:18** =

7) **25:5** =

8) **3:21** =

9) **20:15** =

10) **40:8** =

11) **12:16** =

12) **42:7** =

13) **18:72** =

14) **70:10** =

15) **35:28** =

16) **48:54** =

17) **9:81** =

18) **18:30** =

19) **6:24** =

A ratio compares the size or quantity of two different things. A **proportion** is two or more ratios that equal one another. For example, 10:3 is equal to 20:6.

	Example:	As a Fraction:	Written with Colons: (and in Word Form.)
Ratio:	4 rocks for every tree.	$\dfrac{4 \text{ rocks}}{1 \text{ tree}}$	4:1 (4 to 1)
Proportion:	4 rocks for every tree is equal to 8 rocks for every 2 trees.	$\dfrac{4 \text{ rocks}}{1 \text{ tree}} = \dfrac{8 \text{ rocks}}{2 \text{ trees}}$	4:1::8:2 (4 is to 1 as 8 is to 2)

Directions: *Circle "R" if the example is a ratio and circle "P" if it's a proportion. Then, write the example as a fraction and using colons.*

	Example:	R or P	As a Fraction:	With Colons:
Ex)	6 pens for every 15 pencils is equal to 2 pens for every 5 pencils.	R or **(P)**	$\dfrac{6 \text{ pens}}{15 \text{ pencils}} = \dfrac{2 \text{ pens}}{5 \text{ pencils}}$	**6:15::2:5**
1)	Administer 25 mg per kilogram.	R or P		
2)	25 mg per kilogram is equal to 1,000 mg per 40 kilograms.	R or P		
3)	13 pigs for every 6 goats is equal to 39 pigs for every 18 goats.	R or P		
4)	There are 39 pigs for every 18 goats.	R or P		
5)	Laura earns $30 per hour.	R or P		
6)	$30 per hour is equal to $120 every 4 hours.	R or P		
7)	Every 3 hours Kyle makes $100.	R or P		
8)	The train is driving 90 km per hour.	R or P		
9)	90 km per hour is equal to 450 km every 5 hours.	R or P		
10)	45 boxes for every 3 trucks is equal to 15 boxes per truck.	R or P		

74) Proportions

Sometimes, you may not know all the numbers for a proportion. The unknown numbers are called **unknown terms**. When a proportion has an unknown term, the numbers that are known may be used to calculate the value of the unknown term.

Since proportions are two or more ratios that are equal to one another, you can follow the steps below to find an unknown term.

> **Step 1**: Write the information as a **proportion** in fraction form.
>   (It doesn't matter what letter you use to represent the unknown term.)
> **Step 2**: Identical unit labels cancel each other.
> **Step 3**: Cross multiply and rewrite the problem.
> **Step 4**: Divide to isolate the unknown term.
> **Step 5**: Solve the equation. Doing so will calculate the value of the unknown term.

	Example 1: If Ben picks 12 apples every 3 minutes, how long will it take to pick 80 apples?	**Example 2:** If 10 buckets hold 250 apples, how many buckets are needed to hold 150 apples?
Step 1:	$\dfrac{12 \text{ apples}}{3 \text{ minutes}} = \dfrac{80 \text{ apples}}{m}$	$\dfrac{10 \text{ buckets}}{250 \text{ apples}} = \dfrac{b}{150 \text{ apples}}$
Step 2:	$\dfrac{12 \text{ apples}}{3 \text{ minutes}} = \dfrac{80 \text{ apples}}{m}$	$\dfrac{10 \text{ buckets}}{250 \text{ apples}} = \dfrac{b}{150 \text{ apples}}$
Step 3:	$\dfrac{12}{3 \text{ minutes}} \times \dfrac{80}{m}$	$\dfrac{10 \text{ buckets}}{250} \times \dfrac{b}{150}$
Step 4:	$240 \text{ minutes} = 12 \times m$	$250 \times b = 1{,}500 \text{ buckets}$
Step 5:	$\dfrac{240 \text{ minutes}}{12} = \dfrac{12 \times m}{12}$	$\dfrac{250 \times b}{250} = \dfrac{1{,}500 \text{ buckets}}{250}$
Answer:	$20 \text{ minutes} = m$	$b = 6 \text{ buckets}$

Directions: *Find the value of each unknown term.*

1) $\dfrac{18}{a} = \dfrac{3}{2}$

2) $\dfrac{36}{28} = \dfrac{w}{7}$

3) $14:84::h:30$

4) $3:y::54:36$

5) If Beth bakes 24 cookies every 20 minutes, how many cookies will she bake in 120 minutes?

6) It took Beth 50 minutes to wash 320 plates. At that rate, how many plates can Beth wash in 10 minutes?

75) Proportions

Directions: *Find the value of each unknown term.*

1) $\dfrac{6}{3} = \dfrac{10}{t}$

2) $\dfrac{2}{n} = \dfrac{1}{3}$

3) $\dfrac{10}{5} = \dfrac{u}{7}$

4) $\dfrac{x}{310} = \dfrac{14}{5}$

5) $\dfrac{7}{9} = \dfrac{7}{a}$

6) $\dfrac{21}{45} = \dfrac{d}{15}$

7) $\dfrac{5}{2} = \dfrac{c}{4}$

8) $\dfrac{4}{20} = \dfrac{2}{y}$

9) $\dfrac{26}{e} = \dfrac{13}{21}$

10) $\dfrac{r}{12} = \dfrac{40}{5}$

11) $\dfrac{19}{3} = \dfrac{n}{18}$

12) $\dfrac{1}{w} = \dfrac{8}{64}$

13) $1:9::6:a$

14) $4:5::m:10$

15) $30:c::25:60$

16) $2:1::14:y$

17) $h:75::280:60$

18) $6:10::u:5$

19) $15:w::45:24$

20) $38:18::19:e$

21) $d:18::1:6$

22) $9:78::r:130$

23) $6:a::12:8$

24) $n:50::35:125$

25) If a car travels 120 miles in 3 hours and remains at the same speed, how far will the car travel in 8 hours?

26) A painter is mixing red and yellow paint at a ratio of 3:5. If the painter uses 6 liters of red paint how much yellow paint is needed?

27) The ratio of men to women at the office is 6:7. If there are 30 men, how many women are there?

28) Ben's latte recipe calls for espresso to steamed milk in a 1:4 ratio. If Ben uses 24 ounces of steamed milk, how many ounces of espresso are needed?

29) A cookie recipe requires 5 cups of flour for every 2 cups of sugar. If you want to use 10 cups of sugar, how much flour will you need?

30) A gallon of paint can cover 90 square feet. How many gallons of paint are needed to paint a wall that has 720 square feet?

76) Proportions

Directions: *Find the value of each unknown term.*

1) $\dfrac{6}{2} = \dfrac{c}{5}$

2) $\dfrac{1}{3} = \dfrac{50}{y}$

3) $\dfrac{40}{e} = \dfrac{20}{3}$

4) $\dfrac{r}{56} = \dfrac{38}{28}$

5) $\dfrac{12}{5} = \dfrac{n}{5}$

6) $\dfrac{8}{w} = \dfrac{12}{3}$

7) $\dfrac{70}{10} = \dfrac{7}{t}$

8) $\dfrac{5}{n} = \dfrac{6}{12}$

9) $\dfrac{3}{9} = \dfrac{u}{33}$

10) $\dfrac{x}{2} = \dfrac{50}{10}$

11) $\dfrac{14}{85} = \dfrac{42}{a}$

12) $\dfrac{76}{8} = \dfrac{d}{76}$

13) $1:4::n:16$

14) $2:y::9:18$

15) $45:63::180:r$

16) $x:7::3:21$

17) $8:24::a:6$

18) $10:h::12:6$

19) $c:32::1:8$

20) $50:1::100:m$

21) $30:55::t:22$

22) $9:w::36:8$

23) $25:4::25:e$

24) $d:30::20:100$

25) Becky's garden has 8 rows of beans for every 3 rows of carrots. If her garden has 32 rows of beans, how many rows of carrots are there?

26) A soup needs 6 cups of water for every 5 cups of rice. If you use 20 cups of rice, how many cups of water will you need?

27) The racecar drove 18 laps around the racetrack in 12 minutes. At this rate, how long will it take the racecar to drive 54 laps around the racetrack?

28) Sarah can read 28 emails in 4 minutes. At this rate, how long will it take her to read 64 emails?

29) Tyler wants to buy 4 pizzas for every 9 people at the party. If there are 63 people at the party, how many pizzas should he order?

30) Emma sold muffins to doughnuts at a 6:11 ratio. If Emma sold 48 muffins, how many doughnuts were sold?

77) Proportions

Proportions can be used to convert one **unit of measure** to another. They can even be used to convert from one **system of measure** to another, such as the metric and customary systems. *You may refer to the steps listed on page 74.* (*Equivalencies may not be exact.*)

Customary System Lengths

Name	Foot	Inch
Abbreviation	ft	in
Household Equivalencies	1 ft 12 in	1 in
Metric Equivalencies	30 cm	2.5 cm

Customary System Mass

Name	Pound	Ounce
Abbreviation	lb	oz
Household Equivalencies	1 lb 16 oz	1 oz
Metric Equivalencies	0.45 kg	28 g

	Example 1: A fish is 20 inches long. How many cm is that?	**Example 2:** The fish weighs 4.5 pounds. How many ounces is that?
Step 1:	$\dfrac{20\ inches}{f} = \dfrac{1\ inch}{2.5\ cm}$	$\dfrac{4.5\ lb}{w} = \dfrac{1\ lb}{16\ oz}$
Step 2:	$\dfrac{20\ inches}{f} = \dfrac{1\ inch}{2.5\ cm}$	$\dfrac{4.5\ lb}{w} = \dfrac{1\ lb}{16\ oz}$
Step 3:	$\dfrac{20}{f} \diagup\!\!\!\!\diagdown \dfrac{1}{2.5\ cm}$	$\dfrac{4.5}{w} \diagup\!\!\!\!\diagdown \dfrac{1}{16\ oz}$
Step 4:	$1 \times f = 50\ cm$	$1 \times w = 72\ oz$
Step 5:	$\dfrac{1 \times f}{1} = \dfrac{50\ cm}{1}$	$\dfrac{1 \times w}{1} = \dfrac{72\ oz}{1}$
Answer:	$f = 50\ cm$	$w = 72\ oz$

Directions: *Use proportions to convert each unit of measure.*

1) **7.5 ft =** _____ in

2) **5.5 oz =** _____ g

3) **192 in =** _____ ft

4) **72 kg =** _____ lb

5) **47 in =** _____ cm

6) **38 in =** _____ cm

7) **35 lb =** _____ kg

8) **35 kg =** _____ lb

9) **100 cm =** _____ in

10) **20 oz =** _____ g

11) **9 in =** _____ ft

12) **10 ft =** _____ in

13) **32 oz =** _____ lb

14) **6.5 lb =** _____ oz

15) **70 g =** _____ oz

16) **A puppy weighs 0.4 pounds. How many ounces is that?**

17) **The puppy is 15 cm long. How many inches is that?**

78) Proportions: Multi-Step Conversions

Sometimes, multiple conversions are needed to solve a problem. This is often true when converting units of time. If you want to know how many hours are in two weeks, you probably won't know how many hours there are in a week. But, you will probably remember that there are 7 days in a week and 24 hours in a day. So, this problem may be solved using 2 steps. You can first convert from weeks to days and then from days to hours.

Units of Time

Name	Week	Day	Hour	Minute	Second
Abbreviation	week	day (or d)	hr (or h)	min	sec (or s)
Equivalent Measures	7 days	24 hr	60 min	60 sec	-------

Example: How many hours are in 2 weeks?

	Step 1: Weeks to Days	**Step 2:** Days to Hours
Step 1:	$\dfrac{2 \text{ weeks}}{d} = \dfrac{1 \text{ week}}{7 \text{ days}}$	$\dfrac{14 \text{ days}}{h} = \dfrac{1 \text{ day}}{24 \text{ hours}}$
Step 2:	$\dfrac{2 \text{ weeks}}{d} = \dfrac{1 \text{ week}}{7 \text{ days}}$	$\dfrac{14 \text{ days}}{h} = \dfrac{1 \text{ day}}{24 \text{ hours}}$
Step 3:	$\dfrac{2}{d} \quad \dfrac{1}{7 \text{ days}}$	$\dfrac{14}{h} \quad \dfrac{1}{24 \text{ hours}}$
Step 4:	$1 \times d = 14 \text{ days}$	$1 \times h = 336 \text{ hours}$
Step 5:	$\dfrac{1 \times d}{1} = \dfrac{14 \text{ days}}{1}$	$\dfrac{1 \times h}{1} = \dfrac{336 \text{ hr}}{1}$
Answer:	$d = 14 \text{ days}$	$h = 336 \text{ hours}$

Directions: *Use proportions to convert each unit of measure.*

1) **1.2 days =** _____ s

2) **4.5 weeks =** _____ hr

3) **840 hr =** _____ weeks

4) **1 week =** _____ min

5) **7 hr =** _____ s

6) **5,040 min =** _____ days

7) **6.5 days =** _____ min

8) **5,760 s =** _____ hr

9) **0.1 weeks =** _____ s

10) **345,600 s =** _____ days

11) **20 weeks =** _____ hr

12) **3 days =** _____ s

13) **11 hr =** _____ s

14) **462 hr =** _____ weeks

15) **8 days =** _____ min

16) How many minutes are in 2.75 days?

17) How many seconds are in 3.5 hours?

79) Proportions: Parts of a Total Quantity

You've learned how to calculate the unknown term of a proportion. Sometimes, the proportion is not known though. You may only know the ratio and the total amount. But this information can be used to calculate the unknown term.

Step 1: **Add** the two portions of the ratio together.
Step 2: **Divide** the portion you are solving for by the answer to Step 1.
    This determines the percentage of the ratio for the desired portion.
    (In Example 1, we want to know the percentage of red paint used in the ratio.)
Step 3: **Multiply** the answer of Step 2 (the percentage) with the total amount.

	Example 1: A painter mixed red and blue paint together in a 7:13 ratio. If the painter mixed a total of 3,000 mL of paint, how much red paint was used?	**Example 2:** Emily wants to plant 5 roses for every 11 tulips. If she wants to plant a total of 1,200 flowers, how many tulips should she plant?
Step 1:	7 + 13 = **20**	5 + 11 = **16**
Step 2:	7 ÷ 20 = **0.35** (35%)	11 ÷ 16 = **0.6875** (68.75%)
Step 3:	0.35 × 3,000 mL = **1,050 mL**	0.6875 × 1,200 = **825**
Answer:	**1,050 mL** of red paint was used.	Emily should plant **825 tulips**.

Directions: *Use your knowledge of proportions to calculate a part of the total quantity.*

1) Ava's smoothie shop blends strawberries and blueberries in a 3:5 ratio. If 6,400 mL of smoothie was made, how many mL of blueberries were used?

2) Josh wants to plant 1 oak tree for every 7 pine trees. If he plans to plant a total of 500 trees, how many pine trees will he need?

3) The construction crew mixed gravel and sand in a 3:2 ratio. If 2,000 lb of the mixture was made, how much gravel was used?

4) A recipe calls for sugar and flower to be added at a 2:30 ratio. If a total of 40 kg of the mixture is made, how many kg of flower was used?

5) A scientist mixed bromine and water at an 8:72 ratio. If 250 mL of the solution was created, how much bromine was used?

6) Luke needs to order 9 black shirts for every 16 white shirts. If he wants to order a total of 750 shirts, how many black shirts should he order?

80) Proportions

Determine whether an unknown term of a proportion needs to be calculated, or whether a part of the total quantity needs to be calculated. Then, solve the problem.

1) Cement and sand were mixed using a 2:7 ratio. If 3,000 lb. of cement were used, how many pounds of sand were used?

2) Lily's granola recipe requires walnuts and almonds at a 3:2 ratio. If the total weight of these nuts is 1,200 grams, how many grams of almonds were used?

3) A farmer fed 4 bales of hay for every 25 cows. If there are a total of 500 cows, how many bales of hay were fed?

4) Ryan hit 11 balls for every 15 pitches thrown. If 90 pitches were thrown, how many balls did Ryan hit?

5) Maria planted 11 carrot seeds for 5 bean seeds. If a total of 592 seeds were planted, how many of the seeds were beans?

6) Tina ordered 7 white flowers for every 13 red flowers at her wedding. If a total of 160 flowers were ordered, how many of the flowers were red?

7) When Zoe works out, she lifts weights and runs at a 12/20 ratio. She worked out for a total of 216 minutes this week. How much of that time was spent running?

8) Paul wants to order 5 pizzas for every 12 people invited to his party. If 72 people are invited to the party, how many pizzas should he order?

9) A barista's recipe calls for 25 mL of milk for every 175 mL of coffee. If the barista uses this recipe to fill a mug with 800 mL of liquid, how much of the liquid is coffee?

10) A plane is traveling at a speed of 570 miles per hour. At this rate, how far will the plane fly after 4.5 hours?

11) Water leaking from a pipe is filling 6 buckets of water every 20 minutes. It will take 90 minutes to fix the leak. At this rate, how many buckets of water will be filled before the leak is fixed?

12) Bill's bread recipe calls for 30 mL of salt for every 1970 mL of flour. If Bill wants to make a total of 7,000 mL of the salt and flour mixture. How much salt will he need?

81) Coordinate Planes

A **coordinate plane** has an **X-axis** and **Y-axis**. The X-axis measures distances right and left. The Y-axis measures distances up and down.

Coordinate planes are often called **graphs**. **Coordinates** can be given to name a location on a graph. The value of X is listed first, and the value of Y is listed second. A comma separates both values, and they are enclosed with parentheses (X, Y).

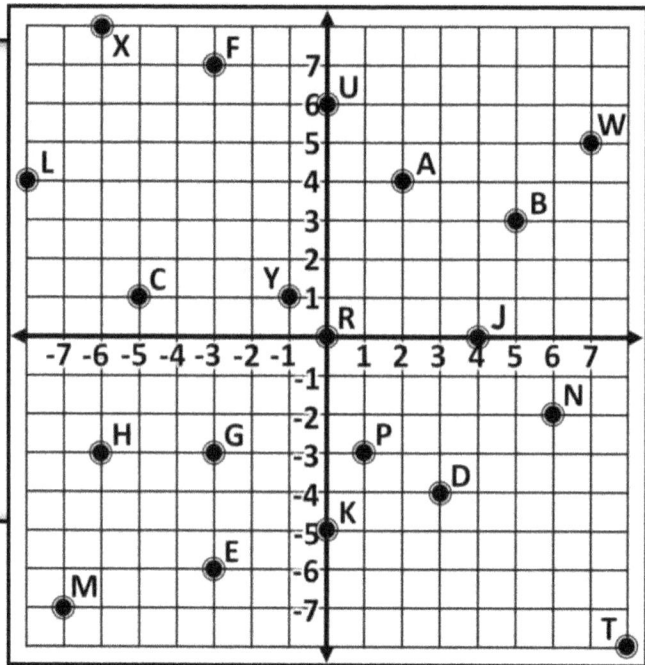

Coordinate Locations

	X	Y	Coordinate
Point A	2	4	(2, 4)
Point C	−5	1	(−5, 1)
Point E	−3	−6	(−3, −6)
Point K	0	−5	(0, −5)

Directions: *Write the coordinate of each point located on the graph above.*

1) **Point B**   5) **Point H**   9) **Point N**   13) **Point U**

2) **Point D**   6) **Point J**   10) **Point P**   14) **Point W**

3) **Point F**   7) **Point L**   11) **Point R**   15) **Point X**

4) **Point G**   8) **Point M**   12) **Point T**   16) **Point Y**

A coordinate can also be used to **plot** a point on a graph. To plot a point, mark a dot at the coordinate's location. The dot is often **labeled** to make it easier to identify.

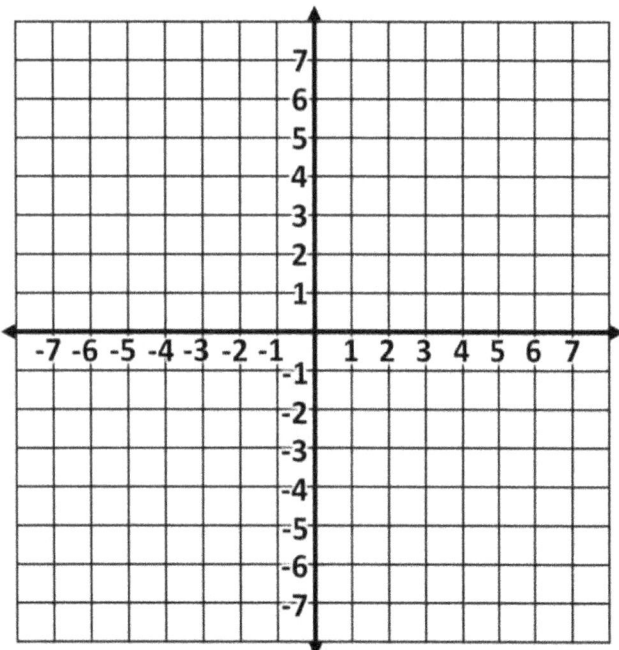

Directions: *Plot and label each point.*

17) A = (1, −6)   27) W = (1, 6)

18) B = (−7, 2)   28) L = (7, 7)

19) C = (2, 0)   29) M = (3, −8)

20) D = (4, 4)   30) E = (−3, 0)

21) U = (−8, −3)   31) J = (−5, 5)

22) F = (8, −3)   32) P = (−3, 3)

23) G = (−5, −8)   33) K = (−4, 7)

24) H = (0, 2)   34) R = (−6, −5)

25) N = (−4, −2)   35) Y = (5, −5)

26) T = (0, −4)   36) X = (1, −1)

82) Coordinate Planes

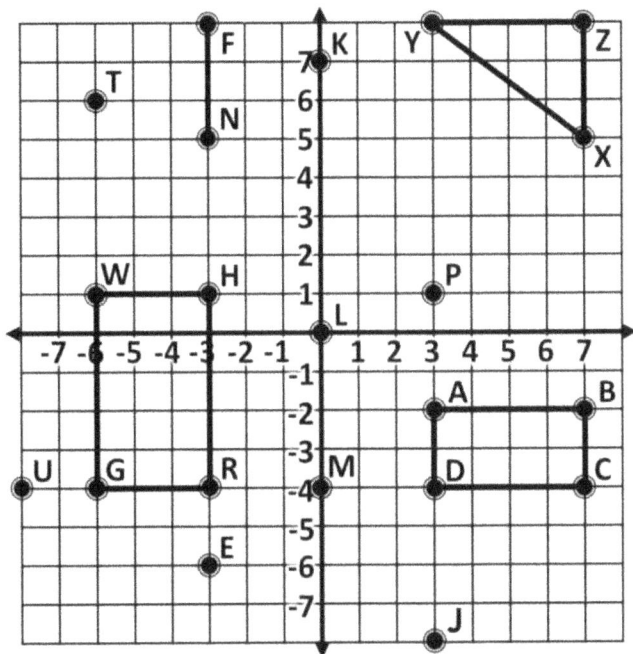

Directions:

Name the point located at each coordinate.

1) **(3, −8)**
2) **(−6, 6)**
3) **(3, −4)**
4) **(−6, 1)**
5) **(3, 1)**
6) **(7, −2)**
7) **(−3, 1)**
8) **(7, 5)**
9) **(−6, −4)**
10) **(7, 8)**

11) **(−3, −4)**
12) **(3, 8)**
13) **(0, 0)**
14) **(−3, −6)**
15) **(3, −2)**
16) **(0, −4)**
17) **(−3, 5)**
18) **(−3, 8)**
19) **(0, 7)**
20) **(7, −4)**

Coordinates can be used to draw lines and measure distances. These lines are usually named using their end points. For example, the line segment that connects Point F and Point N would be called line segment FN.

Some line segments are drawn on the graph above. Other lines are not drawn but can still be described using their end points. For instance, line segment RM is not drawn, but you can determine its location and length based the positions of Point R and Point M.

Measuring Distances	
End Points	**Distance**
Point Y and **Point Z**	4 units
Point G and **Point W**	5 units
Point G and **Point T**	10 units
Point R and **Point M**	3 units

To find the distance between two points, count the number of spaces between the points. For instance, the distance between Point Y and Point Z is 4 units and the distance between Point R and Point M is 3 units.

Directions: *Write the distance between each set of points.*

21) **Point A** and **Point B** =
22) **Point A** and **Point D** =
23) **Point A** and **Point P** =
24) **Point A** and **Point J** =
25) **Point B** and **Point C** =
26) **Point B** and **Point X** =
27) **Point B** and **Point Z** =
28) **Point F** and **Point Z** =
29) **Point L** and **Point M** =
30) **Point K** and **Point M** =

31) **Point Y** and **Point J** =
32) **Point Y** and **Point P** =
33) **Point Y** and **Point Z** =
34) **Point Y** and **Point F** =
35) **Point F** and **Point Z** =
36) **Point F** and **Point N** =
37) **Point F** and **Point E** =
38) **Point W** and **Point H** =
39) **Point W** and **Point P** =
40) **Point W** and **Point T** =

83) Coordinate Planes

Directions:

Fill in the blank to make each statement true.

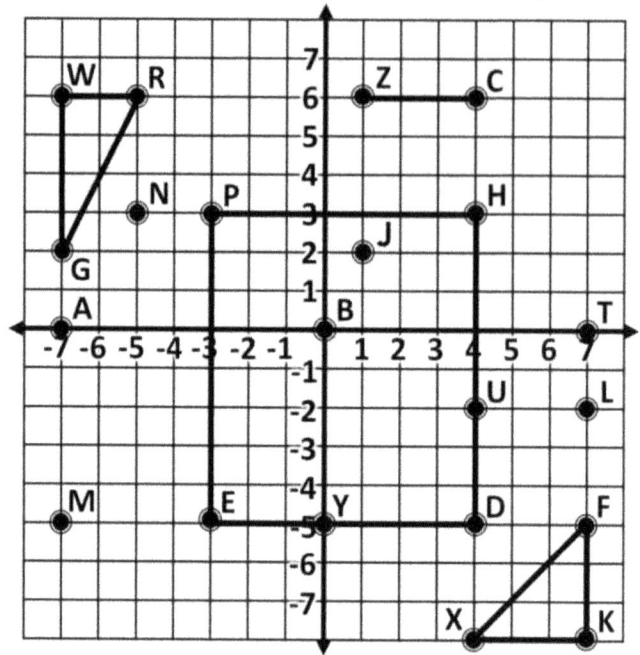

1) Point J is 1 unit right
and 7 units above point ____.

2) Point W is 4 units left
and 3 units above point ____.

3) Point M is 7 units below point ____.

4) Point E is 7 unit left
and 11 units below point ____.

5) Point K is 12 units right
and 14 units below point ____.

6) Point F is ____ units right and ____ units below point H.

7) Point A is ____ units left and ____ units above point U.

8) Point C is ____ units left and ____ units above point U.

9) Point L is ____ units right and ____ units below point W.

10) Point A is ____ units left and ____ units above point Y.

11) The distance from point B to point T is ____ unit(s) less than point R to point C.

12) The distance from point H to point X is ____ unit(s) more than point Z to point J.

13) Line HD is _____ unit(s) longer than line HU.

14) Line ED is _____ unit(s) shorter than line AT.

15) Line MD is _____ unit(s) longer than line XK.

16) Line AG is _____ unit(s) shorter than line FK.

17) If you add the lengths of lines segments UL and GJ, they have a total length of _____ units.

18) If you add the lengths of lines segments BY and TK, they have a total length of _____ units.

19) Points U, L, K, and X can be connected to form a rectangle with a perimeter of _____ units.

20) Rectangle U, L, K, and X has an area of _____ units^2.

21) Points R, C, H, and N can be connected to form a rectangle with a perimeter of _____ units.

22) Rectangle R, C, H, and N has an area of _____ units^2.

23) Points A, T, F, and M can be connected to form a rectangle with a perimeter of _____ units.

24) Rectangle A, T, F, and M has an area of _____ units^2.

84) Coordinate Planes

Pay close attention to the **unit labels** of a coordinate plane. One line does not always equal one unit. One line may equal 10 units. Or, each line may equal 0.5 units, which is the case for the X-axis of the coordinate plane to the right.

Directions: *Name the point located at each coordinate.*

1) **(0, −5)**

2) **(−2, 20)**

3) **(1, 10)**

4) **(1.5, −10)**

5) **(−2.5, 5)**

Directions: *Write the coordinate of each point.*

6) **Point B**

7) **Point F**

8) **Point C**

9) **Point E**

10) **Point D**

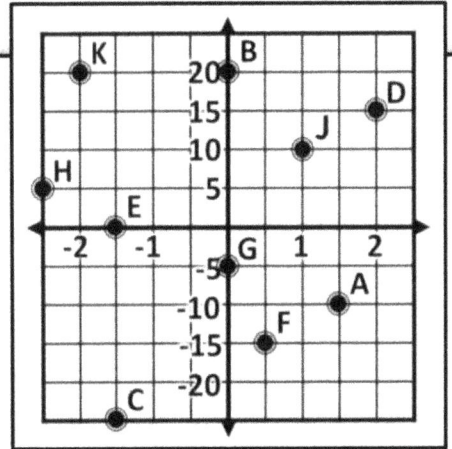

The points on a coordinate plane are not always located on a line. They may be in the spaces between the lines. The exact coordinates of these points may be difficult to determine, but they may be **estimated**. For example, point L is near line 10 of the X-axis and line −20 of the Y-axis. So, the coordinates for point L could be estimated to be (9, −22).

Directions: *Name the point located at each coordinate.*

11) **(5, 40)**

12) **(−17, −57)**

13) **(20, −50)**

14) **(−8, 3)**

15) **(19, 80)**

16) **(−25, 0)**

Directions: *Write an* <u>*estimated*</u> *coordinate for each point.*

17) **Point U**

18) **Point N**

19) **Point Z**

20) **Point R**

21) **Point P**

22) **Point M**

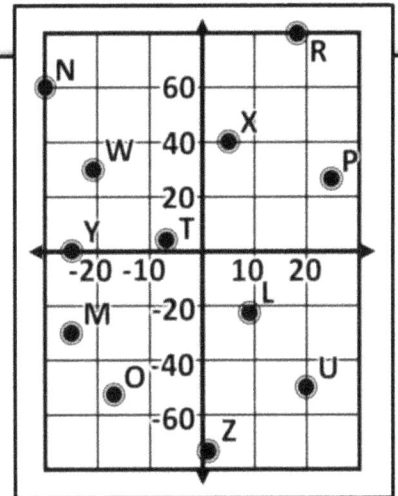

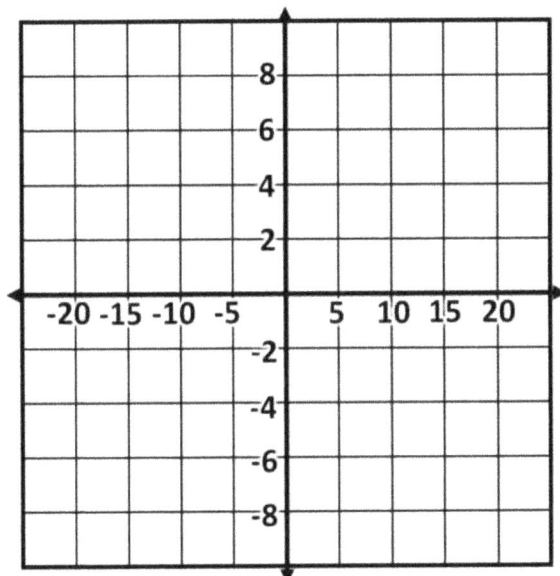

Directions: *Plot and label each point.*

23) **A = (−9, 4)**

24) **P = (8, 0)**

25) **O = (18, −5.8)**

26) **F = (0, −3.9)**

27) **L = (−4.8, 0)**

28) **N = (18.3, 3)**

29) **G = (−15.9, 3)**

30) **R = (−10, −7)**

31) **K = (11, 7)**

32) **E = (23.5, 9)**

33) **C = (−15, −4.4)**

34) **M = (−22.7, 10)**

35) **D = (16.9, −3)**

36) **H = (−24.4, −10)**

37) **X = (0, 6.1)**

38) **Z = (−24.5, −3)**

39) **J = (6, −7.8)**

40) **B = (1.1, 1.7)**

85) Input-Output Tables

An **input-output table** can be used to organize numbers that <u>follow the same rule</u>. The rule is an equation that may be used to calculate unknown numbers for the input-output table. When trying to calculate the value of Y, insert the value of X into the equation. When trying to solve for X, insert the value of Y into the equation.

Example:	Step 1: Insert the value of X.	Step 2: Solve the equation.	Answer: Record your answers.

Example: $X - 2 = Y$

X	Y
-2	
-1	
0	
1	
2	
3	

Step 1: Insert the value of X.

$(-2) - 2 = Y$
$(-1) - 2 = Y$
$0 - 2 = Y$
$1 - 2 = Y$
$2 - 2 = Y$
$3 - 2 = Y$

Step 2: Solve the equation.

$(-2) - 2 = -4$
$(-1) - 2 = -3$
$0 - 2 = -2$
$1 - 2 = -1$
$2 - 2 = 0$
$3 - 2 = 1$

Answer: Record your answers. $X - 2 = Y$

X	Y
-2	-4
-1	-3
0	-2
1	-1
2	0
3	1

Directions: *Fill in the blank spaces to complete each input-output table.*

1) $X + 3 = Y$

X	Y
-3	
-2	
-1	
0	
1	
2	
3	

2) $3X = Y$

X	Y
-3	
-2	
-1	
0	
1	
2	
3	

3) $5 - X = Y$

X	Y
-3	
-2	
-1	
0	
1	
2	
3	

4) $10X + 2 = Y$

X	Y
-3	
-2	
-1	
0	
1	
2	
3	

5) $5X - 10 = Y$

X	Y
-3	
-2	
-1	
0	
1	
2	
3	

6) $X + (-7) = Y$

X	Y
-3	
-2	
-1	
0	
1	
2	
3	

7) $X - (-1) = Y$

X	Y
-3	
-2	
-1	
0	
1	
2	
3	

8) $4X \div 2 = Y$

X	Y
-3	
-2	
-1	
0	
1	
2	
3	

Unstoppable Owl

86) Input-Output Tables

Directions: *Fill in the blank spaces to complete each input-output table.*

1) **X + 5 = Y**

X	Y
−5	
−2	
0	
3	
7	
11	
16	

2) **3 − X = Y**

X	Y
1	
2	
5	
7	
9	
20	
23	

3) **X ÷ 2 = Y**

X	Y
−30	
−24	
−14	
−10	
−2	
4	
12	

4) **1 − 2X = Y**

X	Y
−9	
−8	
−5	
0	
3	
4	
7	

5) **10X ÷ 5 = Y**

X	Y
−10	
−8	
−7	
−6	
−4	
−2	
−1	

6) **−5 + X = Y**

X	Y
−40	
−30	
−15	
−5	
0	
4	
8	

7) **2X = Y**

X	Y
−50	
−20	
−15	
−10	
−3	
1	
7	

8) **X − 8 = Y**

X	Y
2	
3	
8	
10	
16	
21	
30	

9) **X + 10 = Y**

X	Y
−90	
−57	
−32	
−6	
4	
12	
35	

10) **3X − 1 = Y**

X	Y
−10	
−8	
−3	
−1	
0	
2	
5	

11) **−6 + X = Y**

X	Y
−15	
−11	
−6	
3	
5	
9	
14	

12) **10X ÷ 2 = Y**

X	Y
−5	
−4	
−1	
2	
6	
8	
10	

87) Coordinate Plains & Input-Output Tables

It's easy to plot the data from an input-output table on a coordinate plane. Once the value for X and the value for Y are known, you have the coordinates needed to plot a point.

In this section, the input-output tables will have **linear equations**. The word **linear** means straight. When you plot the points for a linear equation it will create a **linear line**, which means it will be a straight line.

Plot the points for each input-output table's linear equation, then draw a straight line through the points to show the equation's linear line.

Example:	**Step 1:** Fill the blanks.	**Step 2:** Plot the points and draw a line.

Example: $X + 1 = Y$

X	Y
−5	
−3	
−2	
0	
1	
4	

Step 1: $X + 1 = Y$

X	Y
−5	−4
−3	−2
−2	−1
0	1
1	2
4	5

Directions: *Complete each input-output table. Then, plot the points and draw the linear line.*

1) $X - 2 = Y$

X	Y
−4	
−3	
−1	
0	
2	
6	

2) $6 - X = Y$

X	Y
−4	
−2	
0	
3	
6	
10	

3) $2X = Y$

X	Y
−4	
−2	
0	
1	
3	
4	

4) $X \div 2 = Y$

X	Y
−10	
−8	
−5	
2	
7	
10	

88) Slope

The slope of a mountain refers to how steep its side is. The **slope** of a line refers to how steep the line is. More specifically, slope <u>measures the vertical change of a line compared to the horizontal change of the line</u>.

Many people like to describe a line's slope as the rise (vertical change) over run (horizontal change). The slope of line AB is 1/2. The slope of line CD is –4/2. But like other fractions, –4/2 can be reduced to –2/1.

Lines are often categorized by their slope.

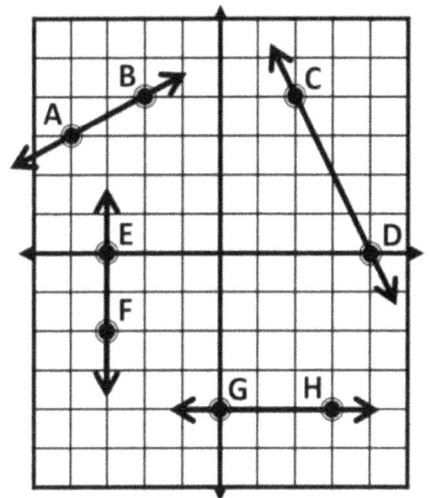

$$\text{Slope} = \frac{\text{Rise}}{\text{Run}}$$

<u>**Positive**</u> lines have a positive (inclining) slope. The value of Y increases as it moves further right. Line AB has a positive slope.

The slope of line AB = $\frac{\text{Rise}}{\text{Run}}$ = $\frac{1}{2}$

<u>**Negative**</u> lines have a negative (declining) slope. The value of Y decreases as it moves right. Line CD has a negative slope.

The slope of line CD = $\frac{\text{Rise}}{\text{Run}}$ = $\frac{-4}{2}$ = $\frac{-2}{1}$

<u>**Vertical**</u> lines have an undefined slope. The line does not move left or right. Line EF is a vertical line.

The slope of line EF = Undefined

<u>**Horizontal**</u> lines have a horizontal (flat) slope. The value of Y does not change as it moves. Line GH has a horizontal slope.

The slope of line GH = $\frac{\text{Rise}}{\text{Run}}$ = $\frac{0}{1}$ = 0

Directions: *Find the slope of each line. Then, indicate whether the slope is positive (P), negative (N), vertical (V), or horizontal (H) by circling the corresponding letter.*

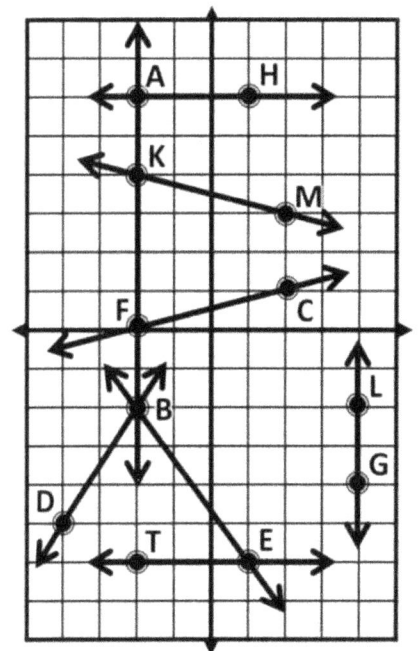

1) **The slope of line AH =**   P   N   V   H

2) **The slope of line AB =**   P   N   V   H

3) **The slope of line KM =**   P   N   V   H

4) **The slope of line FC =**   P   N   V   H

5) **The slope of line DB =**   P   N   V   H

6) **The slope of line BE =**   P   N   V   H

7) **The slope of line LG =**   P   N   V   H

8) **The slope of line TE =**   P   N   V   H

89) Slope and Coordinate Planes

Directions: *Complete each input-output table, plot the line's points, and find the slope.*

1)

X ÷ 5 = Y

X	Y
−20	
−10	
0	
5	
25	

Slope =

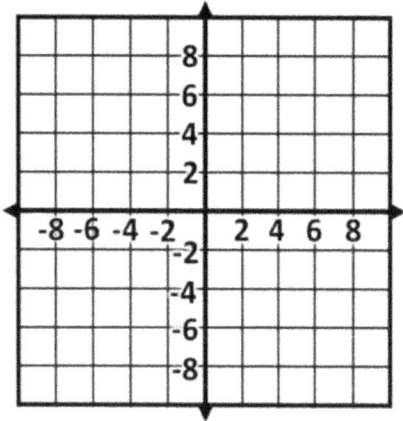

2)

X − 4 = Y

X	Y
−6	
−4	
1	
4	
8	

Slope =

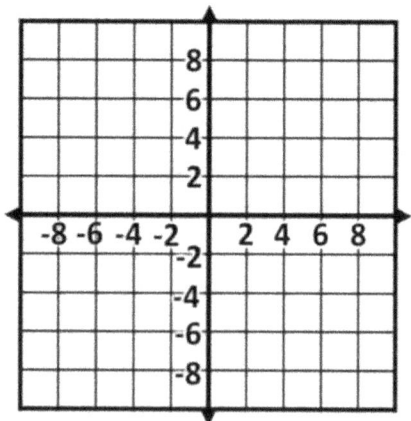

3)

0X + 5 = Y

X	Y
−7	
−3	
0	
5	
8	

Slope =

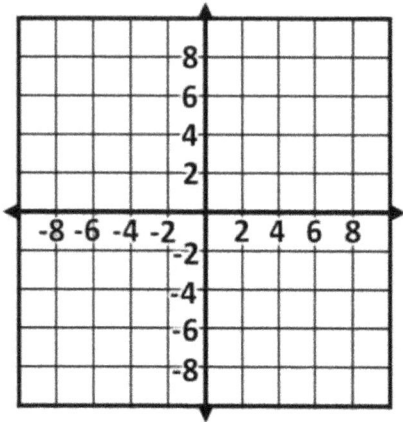

4)

7 + X = Y

X	Y
−10	
−8	
−5	
−1	
1	

Slope =

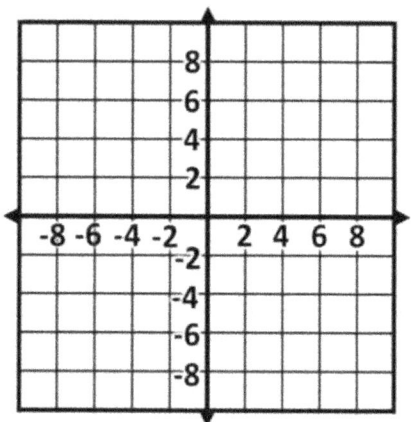

5)

(−1) − X = Y

X	Y
−6	
−3	
−1	
0	
2	

Slope =

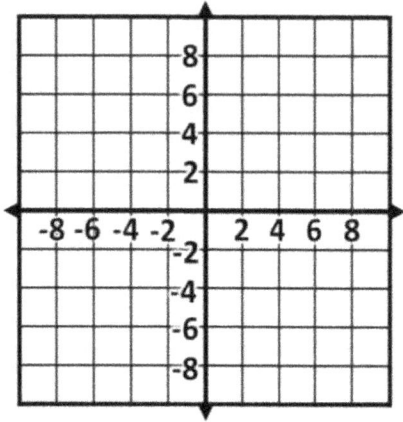

6)

X ÷ (−2) = Y

X	Y
−16	
−10	
0	
6	
14	

Slope =

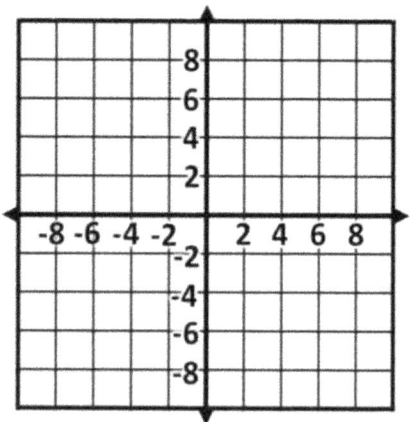

7)

(−2)X = Y

X	Y
−3	
−1	
1	
2	
4	

Slope =

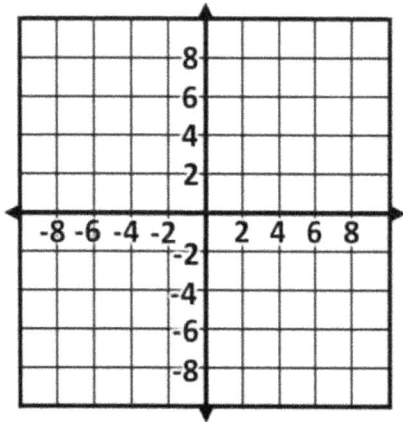

8)

8 − X = Y

X	Y
−1	
2	
5	
7	
10	

Slope =

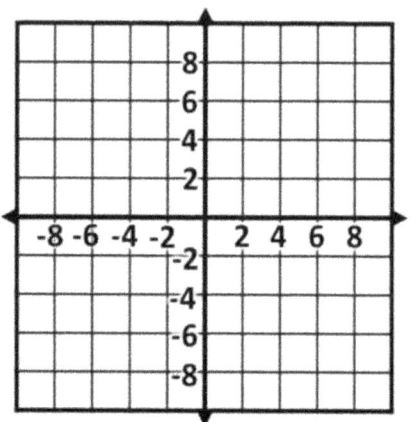

90) Slope Calculations

A line does not need to be graphed to know its slope. When the coordinates of any two points on the line are known, the slope can be calculated. Subtraction can be used to find the change in Y and the change in X between the two points.

$$\text{Slope} = \frac{\text{The change in Y}}{\text{The change in X}} = \frac{Y_2 - Y_1}{X_2 - X_1}$$

The coordinates of the 1st point is (X_1, Y_1).
The coordinates of the 2nd point is (X_2, Y_2).

Example 1: What is the slope of a line that passes through points (−8, 6) and (−2, 2)?

$$\text{Slope} = \frac{Y_2 - Y_1}{X_2 - X_1} = \frac{2 - 6}{-2 - -8} = \frac{-4}{6} = \frac{-2}{3}$$

Does it matter which point is the 1st point and which point is the second point? No, it doesn't. The answer will be the same either way.

Example 2: What is the slope of a line that passes through points (5, −5) and (9, 7)?

If the 1st point is (5, −5) and 2nd point is (9, 7)

$$\text{Slope} = \frac{7 - -5}{9 - 5} = \frac{12}{4} = \frac{3}{1}$$

If the 1st point is (9, 7) and 2nd point is (5, −5)

$$\text{Slope} = \frac{-5 - 7}{5 - 9} = \frac{-12}{-4} = \frac{-3}{-1} = \frac{3}{1}$$

The slope is the same. Either way the slope = 3/1.
−3 divided by −1 is 3/1. (*When dividing a negative by a negative, the answer will be positive.*)

Directions: *Write the slope of the line, in simplest terms, when the line passes through points*:

1) **(−6, 4) and (8, −4)**

2) **(0, 12) and (6, 20)**

3) **(2, −1) and (3, 7)**

4) **(1, 1) and (9, 5)**

5) **(−6, −10) and (−4, −15)**

6) **(−10, 0) and (0, −10)**

7) **(1, 0) and (−3, −8)**

8) **(−20, 4) and (−2, −10)**

9) **(−10, 8) and (5, 8)**

10) **(−26, 4) and (4, 14)**

11) **(−8, −25) and (−1, −4)**

12) **(30, 22) and (16, 8)**

13) **(−15, −10) and (−30, −35)**

14) **(6, 2) and (12, −6)**

15) **(−9, −8) and (−12, 22)**

16) **(−6, 0) and (−6, 5)**

91) Statistics

Data is a collection of facts and information. **Statistics** helps people to understand data. Statistics cannot exist without data, and data is difficult to understand without statistics.

Mean = Average

Step 1: Add all the values in the dataset.
Step 2: Divide the sum by the total number of values.

Data can be interpreted different ways. The **mean** (also known as the **average**) can be calculated by adding all the values in the dataset together. Then, divide the sum by the total number of values.

Name	Cups of Coffee
Mark	11
Sarah	4
Zoe	0
Henry	7
Jane	13
Ryan	4

Example: The chart to the right shows how many cups of coffee each person drank last week. *What is the mean for this dataset?*
(On average, how many cups of coffee did each person drink?)

Step 1: 11 + 4 + 0 + 7 + 13 + 4 = **39**
Step 2: 39 ÷ 6 = **6.5**

Answer: On average, each person drank **6.5 cups of coffee** last week.

Directions: *Calculate the mean for each dataset.* (*Some answers will have a decimal.*)

1) {8, 14, 21, 3, 9}

2) {3, –8, 5, 0, 2, 7, –10}

3) {–13, –1, 4, –5, –10, 8, 1, –6}

4) {51, 23, 67}

5) {2.8, 0.4, 5, 6.2}

6) {0.8, 1.9, 1.7, 7.6}

7) {105, 116, 61}

8) {–7, 12, –5, 4}

9) {–6, –14, –10, 17, 21, 10}

10) {6.3, 7.9, 9.2, 8.6}

11) {5, 14, 26, 33, 41, 12, 6}

12) {–20, 21, –8, 14, –40, 0, 12}

13) {3.6, 15.2, 7, 13.8, 19.3, 7.1, 4}

14) {–7, –5, –3}

15) {–4, 9, 12, –3, 7, 15}

16) {45, 30, 55, 60,70}

17) {2.5, 13.2, 5.3}

18) {12.5, 8.3, 10, 9.1, 11.2, 7.4, 11.5}

19) {8, 9, 10, 6, 7}

20) {120, 80, 70, 85, 150, 100,95}

21) {–0.5, –0.8, –1.3, –2.2, –1.5, –0.7, 0}

22) {25, 13, 7, –14, 22, –31, 11, –9}

23) {2.4, 3.1, –1.2, –0.7, 6.4}

24) {1, 1, 3, 5, 7, 7, 11}

25) **7 people went bowling. Their scores are recorded on the chart below. What is their average score?**

Name	Score
Ben	131
Rose	78
Fay	90
Chris	216
Leo	103
Paula	142
Tina	115

92) Statistics

Median is the "middle" value in a dataset. The median can be found by ordering the values in the dataset from least to greatest. Then, find the number that is in the middle. The middle value is the median.

Median = Middle

Step 1: Order the values <u>least to greatest</u>.
Step 2: Find the middle value.

Example: {6, –11, 8, 10, –5, 0, 4}

Step 1: –11, –5, 0, 4, 6, 8, 10
Step 2: –11, –5, 0, ④ 6, 8, 10
Answer: The median number is **4**.

Directions: *Find the median for each dataset.*

1) {2, –5, 0, –7, 6, 11, 9}

2) {1.1, 0.2, 2.5}

3) {64, –54, 87, 93, 100}

4) {10, 8, –4}

5) {–8, –2, –1, –10, –4}

6) {1.9, 0, –1.2, 9, 1, 0.2, 2.1, –0.6, 1.1, 0.7, 0.3}

7) {3, 22, 15, 8, 12, 1, 13}

8) {–12, 20, –18, 15, 30, 1, 9, 22, –7, 14, –10, 6, 12}

Whenever there is an even amount of numbers in a dataset, there will be two middle numbers. The median will be the value *between* those numbers.

To calculate the value of the median, <u>add the two middle numbers together</u>. Then, <u>divide by 2</u>.

Example: {3, 1, –4, 23, 9, –16, 8, 30}

Step 1: –16, –4, 1, 3, 8, 9, 23, 30
Step 2: –16, –4, 1, ③, ⑧, 9, 23, 30
Step 3: 3 + 8 = **11**
Step 4: 11 ÷ 2 = **5.5**
Answer: The median number is **5.5**.

Directions: *Find the median for each dataset.*

9) {13, 10, 8, 22, 11, 0}

10) {6.4, 4.3, 9.1, 7.5}

11) {–3.7, –0.6, –1.8}

12) {23, 10, 13, 15}

13) {2, 7, 3, –1, 9, 0}

14) {–8, –2, –6, –4}

15) {14, 19, 12, 17, 16}

16) {7, 3, 5, 1, 0, 1, 7}

17) {1.2, 3.4, 5.5, 4}

18) {–15, –22, –9, –30, 25}

19) {3, 6.5, 2, 4.1, 7, 3.9, 3.9, 4.6}

20) {16, –14, 0, –8, 19, 7}

21) {32, 45, 78, 55}

22) {–11, 24, –17, –35, 50, 9, 32, –6, 1, –13}

23) {6, 15, –19, –4, 13, 0}

24) {20, 12, –11, 20}

25) {10.6, 5.4, 11.8, 7.3, –9.7 –4.8}

26) {100, 400, 200, 300}

27) {–79, –85, –92, –110, –50}

28) {930, 1250, 1500}

29) {0.9, 3.1, 1.5, 2.2, 4, 8.9}

93) Statistics

Range is the difference between the largest value and smallest value. Subtract the smallest value from the largest value to find the range.

Range = Difference between the largest and smallest values.
Step 1: Find the smallest and largest values.
Step 2: Subtract the smallest value from the largest.

Directions: *Find the range for each dataset.*

Example: {17, 22, 35, 8, 15, 40}
Step 1: {17, 22, 35, ⑧ 15, ㊵}
Step 2: 40 – 8 = 32
Answer: The range is **32**.

1) {23, 12, 93, 7, 88}

5) {52, 60, 36, 30}

2) {6, –5, 0, –2, 1, –8}

6) {10.5, 6.9, 11.2, 1.8,}

9) {65, 33, 47, 88, 74}

3) {5, 39, 24, –11, –1}

7) {–0.5, –8.7, –1.9, 1}

10) {950, 870, 541, 648}

4) {–10, –20, –7, –4, –9, –13, –18}

8) {20, –14, 19, –8, 17, 2}

11) {–330, –220, –110}

Mode is the value that occurs <u>most</u> often. Some datasets will have a mode, some will have more than one mode, and some datasets will not have a mode (if all the values in a dataset only occur once, there is no mode).

Mode = Most
Step 1: Find the value that occurs the most.

	Step 1:	**Answer:**
Example 1: {–2, –7, 5, 4, 5, 5}	–2, –7, ⑤ 4, ⑤⑤	The mode is 5.
Example 2: {6, –9, 1, –8, 2, 0}	6, –9, 1, –8, 2, 0	There is no mode.
Example 3: {1, 4, 7, 4, 2, 1}	①④ 7, ④ 2, ①	The modes are 1 and 4.

Directions: *Find the mode for each dataset.*

12) {5, 3, 1, 5, 3, 2, 3}

16) {10, 25, 30, 25, 40, 25}

20) {0.5, 1.6, 1.2, 0.5, 1.4}

13) {–1, 0, –5, –5, 0, –7}

17) {–8, 7, –11, 7, 11}

21) {–2.3, 2.3, –1.9, 1}

14) {2, 4, 6, 8, 10, 12}

18) {12, –14, 16, –14, 12}

22) {6, 11, 17, 6, 15, 11, 17, 12}

15) {–3, –5, –7, –5, –9, –8, –4, –2}

19) {36, 82, 35, 5, 36, 57, 5, 7, 82}

23) {–7, –0.8, 0, –1.7, –8, –0.7}

94) Statistics: Mixed Review

You've learned how to calculate the mean, median, range, and mode. Now it's time to calculate each of them for the same data set. Review the previous pages if you need a refresher.

Directions: *Find the mean, median, range, and mode for each dataset.*

	Mean	Median	Range	Mode
1) {–2, –4, –2, –5, –7}	_____	_____	_____	_____
2) {20, 12, 33, 7, 28}	_____	_____	_____	_____
3) {–8, 12, –8, 7, 11}	_____	_____	_____	_____
4) {–3.2, –0.6, –1.7}	_____	_____	_____	_____
5) {8, 2, 1, 10, 4}	_____	_____	_____	_____
6) {–4, –14, –21, –3, –9, –4}	_____	_____	_____	_____
7) {2.5, 3, –1.3, –0.6, 6.4}	_____	_____	_____	_____
8) {30, 30, 25, 75,20}	_____	_____	_____	_____
9) {–10, –13, 8, –8, –13}	_____	_____	_____	_____
10) {21, 13, –7, –14, 22, –21, 13, –7}	_____	_____	_____	_____
11) {6, 3, 5, 5, 4, 2, 3}	_____	_____	_____	_____
12) {23, –14, 13, 15, 17}	_____	_____	_____	_____
13) {215, 310, 301, 200, 150}	_____	_____	_____	_____
14) {20, 14, 19, 18, 20, 20}	_____	_____	_____	_____
15) {30.5, 13.6, 65.9}	_____	_____	_____	_____

95) Statistics Word Problems

Directions: *Indicate if the <u>mean</u>, <u>median</u>, <u>range</u>, or <u>mode</u> is needed. Then, answer the question.*

1) Layla recorded the highest temperature each day for 10 days. The temperatures are listed below. What was the average temperature?

{18, 23, 22, 31, 25, 28, 24, 20, 19, 23}

Mean   Median   Range   Mode

2) By how many degrees did the daily temperature vary?

{18, 23, 22, 31, 25, 28, 24, 20, 19, 23}

Mean   Median   Range   Mode

3) The test scores for Josh's class are listed below. Which test score occurred the most often?

{7, 11, 11, 14, 15, 16, 17, 17, 18, 18
18, 18, 18, 19, 19, 19, 20, 20, 20, 20}

Mean   Median   Range   Mode

4) Josh earned 18 points on the test. He wants to know how his score compares to the class's average score. How many points <u>more or less</u> did Josh earn than the class's average?

{7, 11, 11, 14, 15, 16, 17, 17, 18, 18
18, 18, 18, 19, 19, 19, 20, 20, 20, 20}

Mean   Median   Range   Mode

5) The times for the motorcycle race are recorded below, using minutes and seconds. Once the fastest racer finished, how much time passed before the slowest racer finished?

{7:44, 7:49, 7:51, 7:56, 8:00, 8:03, 8:11}

Mean   Median   Range   Mode

6) After the motorcycle race, the racers parked their motorcycles in a line ordered fastest to slowest finishing times. What was the time of the person parked in the middle of the line?

{7:44, 7:49, 7:51, 7:56, 8:00, 8:03, 8:11}

Mean   Median   Range   Mode

7) Rachel recorded the daily high for a company's stock price for the past 5 days. What was the average daily high for the stock during these 5 days?

{$38.41, $39.22, $38.98, $41.07, $40.79}

Mean   Median   Range   Mode

8) What was the difference between the highest daily high and the lowest daily high for this stock?

{$38.41, $39.22, $38.98, $41.07, $40.79}

Mean   Median   Range   Mode

96) Probability

Probability is the likelihood (chance) of an event. It's not predicting the future. The future is still unknown.

$$\text{Probability} = \frac{\text{Event}}{\text{Outcomes}}$$

Probability is often expressed in fraction form, where the numerator is the number of **events** and the denominator is the total number of possible **outcomes**. For example, if a 6-sided die is rolled, how likely is it to land with the number 5 facing up? There is only one possible event, which is the 5 facing up. But, there are 6 possible outcomes (1, 2, 3, 4, 5, or 6 facing up). So, the probability of a 5 being rolled is 1/6.

What does the fraction mean? The 1/6 means that there is a 1 in 6 chance that a rolled 6-sided die will land with the 5 facing up. (*When possible, these fractions may be reduced.*)

	Events	Outcomes	Probability
Example 1: What is the probability of a 6-sided die rolling a 5?	1 (5)	6 (1, 2, 3, 4, 5, or 6)	$\frac{1}{6}$
Example 2: What is the probability of a 6-sided die rolling an even number?	1 (2, 4, or 6)	6 (1, 2, 3, 4, 5, or 6)	$\frac{3}{6} = \frac{1}{2}$
Example 3: What is the probability of a coin flip landing on heads?	1 (heads)	2 (heads or tails)	$\frac{1}{2}$
Example 4: What is the probability of a 9 being drawn from a deck of cards?	4 (4 nines in a deck.)	52 (52 cards in a deck.)	$\frac{4}{52} = \frac{1}{13}$

Directions: *Find the probability of each event.*
Write the probability as a fraction in <u>simplest form</u>.

                Events   Outcomes   Probability

1) **That a 6-sided die will roll a 3.** _____ _____ _____

2) **That a coin flip will land on tails.** _____ _____ _____

3) **That a 6-sided die will roll a 1 or a 2.** _____ _____ _____

4) **That a queen is drawn from a deck of cards.** _____ _____ _____

5) **That a 6-sided die will roll a 4, 5, or 6.** _____ _____ _____

6) **That a 5 or 8 is drawn from a deck of cards.** _____ _____ _____

7) **That a 6-sided die <u>will not</u> roll a 4.** _____ _____ _____

8) **That a spade is drawn from a deck of cards.** _____ _____ _____

9) **That the 2 of clubs is drawn from a deck of cards.** _____ _____ _____

10) **That a heart <u>is not</u> drawn from a deck of cards.** _____ _____ _____

97) Probability

You've written probability as a fraction. It can be written as a **decimal** and **percentage** too.

- To convert a **fraction to a decimal**, <u>divide the numerator by the denominator.</u>
- To convert a **decimal to a percentage**, <u>multiply by 100.</u>
  (or move the decimal point <u>2 place values to the right</u>.)

Decimals are rounded to the <u>nearest thousandth</u> and percentages to the <u>nearest tenth</u>.

	Fraction	Decimal	Percentage
Example 1: What is the probability of a 6-sided die rolling a 1?	$\frac{1}{6}$	0.167	16.7%
Example 2: What is the probability of a club being drawn from a deck of cards?	$\frac{13}{52} = \frac{1}{4}$	0.25	25%
Example 3: What is the probability of a coin flip landing on tails?	$\frac{1}{2}$	0.5	50%

Directions: *Write the probability as a **fraction** (in <u>simplest form</u>), **decimal** (rounded to the <u>nearest thousandth</u>), and **percentage** (rounded to the <u>nearest tenth</u>).*

	Fraction	Decimal	Percentage
1) That a 6-sided die <u>will not</u> roll a 4.	____	____	____
2) That a 6 or 7 of clubs is drawn from a deck of cards.	____	____	____
3) That a heart <u>is not</u> drawn from a deck of cards.	____	____	____
4) That a 6-sided die will roll a 2 or 4.	____	____	____
5) That a coin flip <u>will not</u> land on tails.	____	____	____
6) That a 2, 3, 4, or 5 is drawn from a deck of cards.	____	____	____

Problems 7-12: *5 red balls, 3 green balls, 8 blue balls, and 4 yellow balls were placed in a bag. If one ball is randomly removed, what is the probability that:*

	Fraction	Decimal	Percentage
7) The ball will be red?	____	____	____
8) The ball will be green?	____	____	____
9) The ball <u>will not</u> be yellow?	____	____	____
10) The ball will be blue or red?	____	____	____
11) The ball <u>will not</u> be green or red?	____	____	____
12) The ball <u>will not</u> be blue?	____	____	____

98) Probability: Multiple Independent Events

To calculate the probability of **multiple independent events**, find the probability of each event separately. Then, multiply those probabilities together. For example, to find the probability of flipping 3 coins and all 3 flips landing on tails, multiply the probability of each event. The probability of each tails is 1/2. So, $1/2 \times 1/2 \times 1/2 = 1/8$.

	Fraction	Decimal	Percentage
Decimals are rounded to the nearest thousandth and percentages to the nearest tenth.			
Example 1: What is the probability of flipping 3 coins and all 3 land on tails?	$\frac{1}{2} \times \frac{1}{2} \times \frac{1}{2} = \frac{1}{8}$	0.125	12.5%
Example 2: What is the probability of rolling a 6-sided die twice and it rolling a 5 the 1st roll and 3 or 4 the 2nd roll?	$\frac{1}{6} \times \frac{1}{3} = \frac{1}{18}$	0.056	5.6%

Directions: *Write the probability as a **fraction** (in simplest form), **decimal** (rounded to the nearest thousandth), and **percentage** (rounded to the nearest tenth).*

	Fraction	Decimal	Percentage
1) That a 6-sided die will roll an even number three times in a row.	___	___	___
2) That a coin flipped twice will land on heads both times.	___	___	___
3) That rolling three 6-sided dice will result in two 6s and an odd number.	___	___	___
4) That a coin flipped 4 times will land on tails each time.	___	___	___
5) That a 6-sided die, when rolled 5 times, will not roll a 1.	___	___	___
6) That a coin flip and roll of a 6-sided die will result in a tails and a 4.	___	___	___
7) That a coin flip and rolling two 6-sided dice will result in a heads and two 3s.	___	___	___
8) That a 6-sided die, when rolled 3 times, will be less than 5 every time.	___	___	___
9) That flipping a coin and drawing a card from a deck of cards will result in a tails and a heart.	___	___	___
10) That 2 coin flips and a 6-sided die roll will result in 2 heads and a number greater than 2.	___	___	___
11) That a 6-sided die, when rolled 4 times, will be larger than 2 every time.	___	___	___
12) That flipping a coin and drawing a card from a deck of cards will result in a heads and a king.	___	___	___

© Libro Studio LLC

99) Probability: Dependent Events

Events can be independent or dependent. **Independent events** do not affect the outcome of each other. If a die is rolled and a coin is flipped, the outcome of the die roll will not affect the outcome of the coin flip, or vice versa.

In some situations, the events do affect each other. In the first example below, there are 10 chocolates in a bag—7 are mint gums and 3 cherry gums. If 2 gums are removed **without replacement**, what are the odds that both will be cherry gums? This example involves dependent events. **Dependent events** will affect the outcome of each other.

When the 1st gum is drawn, there is a 3/10 chance that it will be cherry gum. If the 1st one removed is a cherry gum, there will be a 2/9 chance that the 2nd one removed will be a cherry gum, because there are only 9 gums left in the bag and only 2 of them are cherry. Multiplying these probabilities (3/10 × 2/9) shows that the probability of randomly selecting 2 sticks of gum, without replacement, and getting 2 cherry gums is 6/90, which can be simplified to 1/15.

If the example was selecting 2 sticks of gum from the bag, **with replacement**, and getting 2 cherry gums, the probability would need to be calculated differently. If the 1st gum is replaced after it's drawn, the outcome of the 1st event would not affect the outcome of the 2nd event. They would be independent events. The probability of the 1st gum being a cherry gum is still 3/10, but this time the probability of the 2nd one drawn being a cherry gum would be 3/10 as well. Multiplying these probabilities (3/10 × 3/10) shows the probability of drawing a cherry gum both times would be 6/100, which can be simplified to 3/50.

Scenario: There are 10 sticks of gum in a bag. 7 are mint gums and 3 cherry gums.

Example 1: If 2 gums are randomly removed <u>without</u> replacement, what is the probability that both will be cherry gum?

1st Event		2nd Event		Probability		
$\frac{3}{10}$	×	$\frac{2}{9}$	=	$\frac{6}{90}$	=	$\frac{1}{15}$

Example 2: If 3 sticks are randomly removed <u>without</u> replacement, what is the probability that all 3 will be mint gum?

1st Event		2nd Event		3rd Event		Probability		
$\frac{7}{10}$	×	$\frac{6}{9}$	×	$\frac{5}{8}$	=	$\frac{210}{720}$	=	$\frac{7}{24}$

Directions: *Write the probability as a **fraction** (in <u>simplest form</u>).*

1) **In the scenario above, if 2 gums are randomly removed <u>without</u> replacement, what is the probability that both will be mint gums?**

2) **In the scenario above, if 3 gums are randomly removed <u>without</u> replacement, what is the probability that all 3 will be cherry gums?**

3) **In the scenario above, if 2 gums are randomly removed <u>without</u> replacement, what is the probability that the 1st gum will be mint and the 2nd gum will be cherry?**

4) **2 cards are drawn from a deck of cards, <u>without</u> replacement. What is the probability that they are both 9s?**

5) **2 cards are drawn from a deck of cards, <u>without</u> replacement. What is the probability that they both <u>are not</u> hearts?**

6) **3 cards are drawn from a deck of cards, <u>without</u> replacement. What is the probability that all 3 will be either a king or queen?**

100) Probability: Dependent and Independent Events

Directions: *Write the probability as a **fraction** (in simplest form).*

1) 2 cards are drawn from a deck of cards, <u>with</u> replacement.
   What is the probability that they are both 10s?
2) 3 cards are drawn from a deck of cards, <u>without</u> replacement.
   What is the probability that all 3 will be diamonds?
3) 4 cards are drawn from a deck of cards, <u>with</u> replacement.
   What is the probability that all 4 <u>will not</u> be a king or queen?

Problems 4-10 Scenario: <u>There are 5 blue balls, 8 red balls and 2 white balls in a bag.</u>

4) If 3 balls are randomly removed, <u>without</u> replacement, what is the probability that they are all blue balls?
5) If 2 balls are randomly removed, <u>without</u> replacement, what is the probability that they are both white balls?
6) If 2 balls are randomly removed, <u>with</u> replacement, what is the probability that they are both white balls?
7) If 3 balls are randomly removed, <u>with</u> replacement, what is the probability that they are all red balls?
8) If 3 balls are randomly removed, <u>without</u> replacement, what is the probability that the 1st ball is a white ball and the next 2 balls are blue balls?
9) If 5 balls are randomly removed, <u>without</u> replacement, what is the probability that they are all red balls?
10) If 2 balls are randomly removed, <u>with</u> replacement, what is the probability that the 1st ball is red and the 2nd ball is white?

Problems 11-18 Scenario: <u>There are 11 black beads, 6 clear beads and 3 gray beads in a bag.</u>

11) If 3 beads are randomly removed, <u>with</u> replacement, what is the probability that they are all gray beads?
12) If 4 beads are randomly removed, <u>without</u> replacement, what is the probability that they are all clear beads?
13) If 2 beads are randomly removed, <u>without</u> replacement, what is the probability that they are all black beads?
14) If 2 beads are randomly removed, <u>with</u> replacement, what is the probability that they are all black beads?
15) If 3 beads are randomly removed, <u>without</u> replacement, what is the probability that the 1st bead is black and the next 2 beads are clear?
16) If 6 beads are randomly removed, <u>with</u> replacement, what is the probability that they are all clear beads?
17) If 6 beads are randomly removed, <u>without</u> replacement, what is the probability that they are all clear beads?
18) If 2 beads are randomly removed, <u>without</u> replacement, what is the probability that the 1st bead is gray and the 2nd bead is black?

101) Answer Key

Page 1:
1) −73 2) −97 3) −44 4) −160 5) −9 6) −314 7) −68
8) −51 9) −1237 10) Negative 9 11) Negative 45
12) Negative 258 13) Negative 329 14) Negative 78
15) Negative 8 16) Negative 2041 17) Negative 546
18) Negative 50 19) −25 20) −85

Page 2:
1) −3, −2, 1 2) −9, −8, −1 3) −2, 3, 6 4) −7, −4, −2, 0
5) −1, 2, 4, 6 6) −6, −4, −3 7) −9, −7, −4, −2

Page 3:
1) > 2) > 3) < 4) < 5) < 6) < 7) < 8) > 9) >
10) < 11) > 12) > 13) > 14) < 15) < 16) < 17) >
18) > 19) > 20) > 21) < 22) −59, −57, −55, −53
23) −2, −1, 0, 3 24) −31, −30, −27, −24

Page 4:
1) −10, −8, −2 2) −5, −1, 0 3) −12, −9, −6
4) −13, −3, 0 5) −4, −3, 2 6) −9, 0, 9 7) −7, −1, 0
8) −13, −6, 5 9) −29, −5, 18 10) −23, −8, −4
11) −8, −5, −3, −1 12) −7, −4, 0, 3
13) −10, −9, 9, 10 14) −8, −6, 0, 2
15) −25, −21, −17, −11 16) −44, −32, −29, −14
17) −20, −19, −13, 6 18) −81, −72, −54, 60
19) −41, −27, −18, 34 20) −75, −67, −59, −43
21) −33, −20, −12, 41 22) −80, −74, 29, 45
23) −59, −40, −39, 0 24) −58, −54, −43, 52
25) −94, −82, −40, 61 26) −300, −99, −78, 84
27) −80, −49, −15, 0 28) −100, −81, −73, 99
29) −91, −90, −88, −76 30) −51, −45, −33, −27

Page 5:
1) 2 2) 9 3) 15 4) 27 5) 19 6) 32 7) 11 8) 11 9) 5 10) 29
11) 19 12) 25 13) 64 14) 153 15) 241 16) 85 17) 98
18) 40 19) 66 20) 321 21) 8 22) 33 23) 64 24) 129
25) 270 26) 815 27) 362 28) 714 29) 608 30) 1,000

Page 6:
1) −11 2) −5 3) −7 4) −6 5) −14 6) −15 7) 3 8) 9
9) −1 10) 8 11) −7 12) −3 13) −7 14) −10 15) 5 16) −15
17) −2 18) −13 19) 14 20) 9 21) −6 22) 8 23) −21
24) 10 25) −19 26) 15 27) −1 28) −25 29) 0 30) −12

Page 7:
1) −50 2) −34 3) −38 4) −32 5) 27 6) −40 7) −27
8) 15 9) −54 10) −8 11) −45 12) −20 13) −19 14) 60
15) 25 16) −49 17) 17 18) −24 19) −44 20) −60 21) −96

Page 8:
1) −784 2) −77 3) −215 4) −324 5) 200 6) −548
7) 112 8) 186 9) 164 10) 300 11) −259 12) 192
13) −198 14) 307 15) −100 16) −121 17) 18
18) −300 19) −80 20) 52 21) $-270 + $1,100 = $830

Page 9:
1) −1 2) −8 3) 8 4) −7 5) −9 6) −1 7) −16 8) 9 9) −18
10) 15 11) −18 12) −13 13) 15 14) −1 15) −4 16) 0 17) −5
18) 10 19) −21 20) −1 21) 4 22) −17 23) −4 24) 20

Page 10:
1) −6 2) 12 3) −21 4) −14 5) 11 6) −19 7) 15 8) 14
9) 78 10) 36 11) 88 12) −38 13) 0 14) −80 15) 53
16) 68 17) 88 18) −54 19) −70 20) −56 21) 1 22) −77
23) 90 24) 30 25) −50 26) 43 27) −65 28) −89 29) 38
30) −87 31) −665 32) −120 33) −334 34) 51 35) −395
36) 468 37) −1,082 38) −107 39) 2,501 40) 990

Page 11:
1) −2 2) 10 3) 14 4) −4 5) 14 6) −5 7) −13 8) 24
9) 26 10) −38 11) −59 12) −10 13) −75 14) −76 15) −66
16) −78 17) −35 18) −48 19) −65 20) −63 21) −63 22) 8
23) −48 24) −18 25) −2 26) −71 27) 34 28) 70 29) −1

Page 11 Continued:
30) 47 31) −24 32) −41 33) −30 34) 37 35) −25 36) −78
37) 60 38) −1 39) −29 40) 7 41) 50 42) −54 43) −107
44) 180 45) −70 46) −65 47) −40 48) −60

Page 12:
1) 12 2) −48 3) −20 4) 42 5) 81 6) −56 7) 8 8) 0
9) −24 10) −40 11) −12 12) −10 13) 30 14) −21
15) −18 16) 32 17) −27 18) −12 19) 16 20) −48
21) 49 22) −11 23) 72 24) −14 25) −24 26) −22
27) −55 28) 60 29) 36 30) 0

Page 13:
1) 48 2) −36 3) −44 4) −20 5) −45 6) 8 7) 0
8) −36 9) −21 10) −20 11) 42 12) 27 13) −22
14) 60 15) 99 16) −10 17) 33 18) −72 19) 0
20) 72 21) −12 22) −36 23) 56 24) −20 25) −100
26) 44 27) −45 28) −16 29) 18 30) −25 31) −6
32) 42 33) −72 34) −30 35) 54 36) 55 37) −24
38) 60 39) 48 40) −8 41) 0 42) −9 43) 16 44) −21
45) −24 46) −49 47) −72 48) 132

Page 14:
1) −1 2) −4 3) 3 4) 4 5) −5 6) −7 7) −3 8) 7 9) −2
10) −9 11) 11 12) −1 13) −6 14) 12 15) −11 16) 2
17) −8 18) 4 19) −4 20) −7 21) −6 22) 10 23) 1
24) −7 25) −7 26) 7 27) −12 28) −6 29) 9 30) 8

Page 15:
1) −5 2) 9 3) −4 4) −7 5) −5 6) −10 7) 10 8) 2
9) 9 10) −5 11) −7 12) 10 13) 11 14) −7 15) 8 16) 8
17) 5 18) −11 19) −10 20) 9 21) −2 22) −7 23) 5
24) −5 25) −2 26) −5 27) 3 28) 7 29) 9 30) −11
31) −9 32) 9 33) −5 34) −10 35) 12 36) 12 37) −1
38) 5 39) 4 40) −11 41) 11 42) −12 43) −9 44) −11
45) −10 46) 12 47) 11 48) −10

Page 16:
1) 50 2) −70 3) −5 4) −6 5) −36 6) 16 7) 10 8) −25
9) −28 10) −15 11) −14 12) −16 13) 18 14) −4 15) −24
16) −13 17) 10 18) −20 19) −6 20) 12 21) −13
22) −17 23) 99 24) −10 25) −8 26) 7 27) 9 28) 16
29) −24 30) 5 31) −14 32) −3 33) 33 34) −72 35) 2
36) −1 37) 0 38) −5 39) −1 40) 8 41) −1 42) −9
43) −10 44) 28 45) 2 46) −49 47) −1 48) −8

Page 17:
1) −14 2) −50 3) −44 4) 12 5) −9 6) −39 7) 12
8) −3 9) 4 10) −4 11) −118 12) 8 13) −60 14) −12
15) −36 16) −1 17) 1 18) −60 19) 0 20) 20 21) 6
22) 4 23) 9 24) −2 25) 0 26) −7 27) −56 28) −4
29) 7 30) −64 31) −5 32) −2 33) 7 34) −36 35) −10
36) 12 37) −5 38) 40 39) 7 40) 8 41) 0 42) −44
43) −74 44) −66 45) −6 46) 3 47) −60 48) −3

Page 18:
1) 3 2) −44 3) −31 4) −72 5) −16 6) −105 7) 12
8) −274 9) −8 10) −30 11) −15 12) 15 13) −3 14) 2
15) −44 16) 12 17) −10 18) −36 19) 0 20) −17 21) −3
22) −37 23) 2 24) 5 25) −72 26) 77 27) −26 28) 43
29) 45 30) −1 31) −8 32) 7 33) 6 34) −60 35) 72
36) 20 37) −9 38) −17 39) −16 40) −8 41) −9 42) −63
43) 20 44) −4 45) −9 46) −20 47) −50 48) 99

Page 19:
1) −56 + 8 = −48 feet   2) −7 −11 = −18 degrees
3) −8 × 4 = $-32   4) $60 − $200 = $-140
5) −8 + 12 = 4 degrees   6) $-300 + $500 = $200
7) 9 × $-8 = $-72   8) −30 ÷ 2 = −15 seconds

Page 20:
1) $-45 + $50 = $5   2) −250 + −25 = −275 meters
3) $8 + $-15 = $-7   4) $-12,000 ÷ 3 = $-4000
5) −80 + 100 = 20 points 6) −32 + 20 = −12degrees
7) $200 + $-10 = $190   8) $-2,000 × 4 = $-8,000

Page 21:
1) $8^2 = 8 \times 8$
2) $9^7 = 9 \times 9 \times 9 \times 9 \times 9 \times 9 \times 9$
3) $2^4 = 2 \times 2 \times 2 \times 2$
4) $10^5 = 10 \times 10 \times 10 \times 10 \times 10$
5) one to the eighth power
  $= 1 \times 1 \times 1 \times 1 \times 1 \times 1 \times 1 \times 1$
6) four to the fourth power
  $= 4 \times 4 \times 4 \times 4$
7) eleven to the third power
  $= 11 \times 11 \times 11$
8) seven to the fifth power
  $= 7 \times 7 \times 7 \times 7 \times 7$
9) two to the second power = 2^2
10) twelve to the sixth power = 12^6
11) three to the tenth power = 3^{10}
12) six to the third power = 6^3

Page 22:
1) 1 2) 1 3) 4 4) 16 5) 32 6) 128 7) 256
8) 1,024 9) 2,048 10) 4,096 11) 9 12) 27
13) 243 14) 6,561 15) 4 16) 64 17) 4,096
18) 125 19) 625 20) 3,125 21) 1,296 22) 343
23) 64 24) 729 25) 121 26) 144 27) 400
28) 8,000 29) 10,000 30) 1,000,000 31) 10
32) 100 33) 1,000 34) 10,000 35) 100,000
36) 1,000,000 37) 10,000,000
38) 100,000,000 39) 1,000,000,000
40) 10,000,000,000
41) The exponent tells how many zeros are added after the 1
42) The pattern helps to calculate powers of 10 easily by simply knowing the exponent.

Page 23:
1) 500 2) 900,000 3) 6,000
4) 7,000,000,000,000
5) 400,000,000,000,000 6) 12,000
7) 20,000,000,000 8) 300,000,000
9) 27,000,000 10) 49,000,000,000
11) 10^4 12) 10^8 13) 10^{10} 14) 10^3 15) 10^3
16) 10^4 17) 10^6 18) 10^3 19) 10^3 20) 10^4

Page 24:
1) N 2) Y 3) Y 4) Y 5) N 6) N 7) N 8) N
9) Y 10) N 11) Y 12) N 13) 95,000
14) 4,703 15) 20,000,000 16) 825,000
17) 330 18) 5,000,000,000,000,000
19) 102,490,000,000,000,000,000
20) 716,500,000,000,000,000
21) 600,170,000 22) 9,000,000,000
23) 52,800,000 24) 16,000,000,000
25) 4,788,310

Page 25:
1) 35,000 2) 77 3) 2,000,000,000
4) 890,000 5) 90,000,000 6) 117,000,000
7) 1,257 8) 5,280,000
9) 16,000,000,000 10) 478.831
11) 9.7×10^8 12) 5.6×10^3 13) 3.14×10^5
14) 7.315×10^7 15) 1.5×10^4
16) 6.05×10^8 17) 8×10^7 18) 3.21×10^6
19) 4.79×10^3 20) 2.2×10^2 21) 6.782×10^7
22) 5×10^6 23) 8.47×10^9 24) 4.35×10^{13}
25) 3×10^{17} 26) 1.5×10^{11} 27) 8.202×10^8
28) 7.1×10^{19} 29) 4.503×10^{11}
30) 9×10^{20} 31) 5.86×10^7 32) 2.5×10^{14}
33) 4.59×10^{10} 34) 7.5×10^{23}

102) Answer Key

Page 26:

1) $1 \div 5 \div 5 \div 5 \div 5 \div 5 = 1/5^5$
2) $1 \div 12 \div 12 \div 12 \div 12 = 1/12^4$
3) $1 \div 2 \div 2 \div 2 \div 2 \div 2 = 1/2^5$
4) $1 \div 5 \div 5 \div 5 = 1/5^3$
5) $4^{-5} = 1/4^5$ 6) $15^{-2} = 1/15^2$ 7) $2^{-8} = 1/2^8$
8) $6^{-6} = 1/6^6$ 9) $7^{-4} = 1 \div 7 \div 7 \div 7 \div 7$
10) $30^{-3} = 1 \div 30 \div 30 \div 30$
11) $8^{-6} = 1 \div 8 \div 8 \div 8 \div 8 \div 8 \div 8$
12) $9^{-5} = 1 \div 9 \div 9 \div 9 \div 9 \div 9$

Page 27:

1) 5,764,801 2) 1,771,561 3) 2,097,152 4) 4,913
5) 16,384 6) 194,481 7) 59,049 8) 59,049
9) 2,048 10) 46,656 11) 1,048,576 12) 7,311,616
13) 531,441 14) 15,625 15) 1,048,576 16) 0.0016
17) 0.125 18) 0.0004 19) 0.015625 20) 0.0025
21) 0.03125 22) 0.000001 23) 0.0625
24) 0.000008 25) 0.000125 26) 0.000625
27) 0.0625 28) 0.015625 29) 0.0001 30) 0.0001
31) 0.1 32) 0.01 33) 0.001 34) 0.0001 35) 0.00001
36) 0.000001 37) 0.0000001 38) 0.00000001
39) Each time the negative exponent increases by 1, the decimal moves one place left.
40) It helps to calculate quickly and accurately by understanding how decimals moves.

Page 28:

1) 0.00006 2) 0.0000015 3) 0.000217
4) 0.000000809 5) 0.9 6) 0.00006
7) 0.000000001501 8) 0.0000000008 9) 0.00528
10) 0.000000016 11) 0.07731 12) 0.000000001501
13) 6×10^{-5} 14) 5.2×10^{-3} 15) 7.316×10^{-5}
16) 8×10^{-9} 17) 3.94×10^{-6} 18) 1.5×10^{-4}
19) 5.6×10^{-2} 20) 8×10^{-5} 21) 2.9×10^{-8}
22) 7×10^{-15} 23) 4.4×10^{-11} 24) 9.32×10^{-6}

Page 29:

1) 0.000081 2) 700,000,000 3) 120
4) 0.000000135 5) 7,700 6) 0.000314
7) 0.0000000015 8) 22,000,000,000
9) 0.00000259 10) 0.85 11) 0.0000038
12) 0.0071 13) 0.0125 14) 90,000,000
15) 0.00000000011 16) 0.000000030247
17) 0.00000000006 18) 2,000,000,000,000
19) 3.5×10^8 20) 5.72×10^{-4} 21) 9.8×10^{-7}
22) 1.1×10^5 23) 7.7×10^{-3} 24) 3×10^{-2}
25) 1.21×10^{-5} 26) 5.12×10^7 27) 4×10^7
28) 3.14×10^{11} 29) 4×10^{-9} 30) 6.5×10^9
31) 6.6×10^{-14} 32) 7.18×10^{-18} 33) 4.33×10^{-7}
34) 8.5×10^{14} 35) 1.9×10^{-15} 36) 4.3×10^{17}
37) 0.000000025 meters 38) 9.3×10^7 miles
39) 8.5×10^{-6} g/mL 40) 300,000,000 m/s

Page 30:

1) 8^2 2) 40^2 3) $6,000^2$ 4) 5 5) 9 6) 2 7) 7 8) 10
9) 8 10) 6 11) 12 12) 20 13) 100 14) 60 15) 30

Page 31:

1) 15 2) 18 3) 29 4) 28 5) 25 6) 37 7) 32 8) 35
9) 39 10) 22 11) 33 12) 34 13) 26 14) 50 15) 36
16) 3.46 17) 5.20 18) 6.93 19) 2.45 20) 4.24
21) 10.39 22) 8.94 23) 5.66 24) 10.58 25) 9.95
26) 8.66 27) 7.07 28) 12.12 29) 12.73 30) 22.36

Page 32:

1) 3 2) 2.83 3) 2.63 4) 4.33 5) 3.16 6) 0.22
7) 3.10 8) 3.13 9) 2.52 10) 2.03 11) 3 12) 2
13) 2.51 14) 3.99 15) 3.31 16) 6 17) 2.52 18) 0.2
19) 8.32 20) 4.49 21) 2 22) 0.36 23) 4.45 24) 2.51
25) 6 26) 0.4 27) 2.87 28) 6 29) 4 30) 2.56

Page 33:

1) $(-8)^3 = (-8) \times (-8) \times (-8)$
2) $(-10)^4 = (-10) \times (-10) \times (-10) \times (-10)$
3) $(-20)^2 = (-20) \times (-20)$
4) $(-3)^6 = (-3) \times (-3) \times (-3) \times (-3) \times (-3) \times (-3)$
5) Negative four to the third power $= (-4) \times (-4) \times (-4)$
6) Negative five to the fifth power
   $= (-5) \times (-5) \times (-5) \times (-5) \times (-5)$
7) Negative seven to the second power $= (-7) \times (-7)$
8) Negative eleven to the fourth power
   $= (-11) \times (-11) \times (-11) \times (-11)$
9) Negative six to the seventh power $= (-6)^7$
10) Negative one hundred to the fourth power $= (-100)^4$
11) Negative fifteen to the sixth power $= (-15)^6$
12) Negative three to the third power $= (-3)^3$

Page 34:

1) 4 2) −8 3) 16 4) −32 5) 64 6) −128 7) 256 8) −512
9) 1,024 10) −2,048 11) 9 12) −27 13) 81 14) −243
15) 729 16) 16 17) −64 18) 256 19) −1,024 20) 4,096
21) 25 22) −125 23) 625 24) −3,125 25) 36 26) −216
27) 100 28) −1,000 29) 10,000 30) −100,000 31) If the exponent is an odd number the value is negative, and if it is an even exponent, the value is positive.
32) N 33) P 34) N 35) N 36) P 37) N 38) P 39) P 40) P 41) N
42) P 43) N 44) N 45) P 46) N 47) P 48) P 49) N 50) N

Page 35:

1) $9 - 7 + 2$
   $2 + 2$
   4
2) $20 \div 5 \times 2$
   4×2
   8
3) $15 - 12 + 2$
   $3 + 2$
   5
4) $4 \times 7 - 3$
   $28 - 3$
   25
5) $30 - 6 \div 3$
   $30 - 2$
   28
6) $6 + 4 \times 2$
   $6 + 8$
   14
7) $12 \div 2 \times 3$
   6×3
   18
8) $22 - 2 \times 2$
   $22 - 4$
   18

Page 36:

1) $18 \div 2 + 3$
   $9 + 3$
   12
2) $6 + 4 \div 2$
   $6 + 2$
   8
3) $10 - 5 \times 2$
   $10 - 10$
   0
4) $36 \div 9 - 3$
   $4 - 3$
   1
5) $21 + 7 \times 2$
   $21 + 14$
   35
6) $40 + 10 \times 6$
   $40 + 60$
   100
7) $30 \div 6 - 1$
   $5 - 1$
   4
8) $44 - 4 \times 5$
   $44 - 20$
   24
9) $8 \times 9 - 7$
   $72 - 7$
   65
10) $100 + 50 \times 2$
    $100 + 100$
    200
11) $72 - 6 \times 6$
    $72 - 36$
    36
12) $9 + 13 \times 3$
    $9 + 39$
    48
13) $30 \div 15 + 2$
    $2 + 2$
    4
14) $49 + 7 \times 3$
    $49 + 21$
    70
15) $75 \div 15 \times 2$
    5×2
    10
16) $56 + 11 \times 8$
    $56 + 88$
    144
17) $60 + 45 \div 15$
    $60 + 3$
    63
18) $26 + 17 \times 4$
    $26 + 68$
    94
19) $245 - 10 \times 5$
    $245 - 50$
    195
20) $12 \times 8 - 6$
    $96 - 6$
    90
21) $45 \div 5 + 3$
    $9 + 3$
    12
22) $87 - 45 + 60$
    $42 + 60$
    102
23) $40 \div 5 \times 10$
    8×10
    80
24) $24 + 48 \div 12$
    $24 + 4$
    28
25) $66 \div 11 - 6$
    $6 - 6$
    0
26) $29 + 3 \times 8$
    $29 + 24$
    53
27) $90 \div 9 - 8$
    $10 - 8$
    2
28) $70 \times 10 - 7$
    $700 - 7$
    693
29) $90 \div 5 \times 4$
    18×4
    72
30) $615 + 300 \div 30$
    $615 + 10$
    625

Page 37:

1) $12 \div 2 - 2 \times 0$
   $6 - 2 \times 0$
   $6 - 0$
   6
2) $26 + 5 - 12 \times 2$
   $26 + 5 - 24$
   $31 - 24$
   7
3) $9 \times 8 - 15 \div 5$
   $72 - 15 \div 5$
   $72 - 3$
   69
4) $80 \div 8 + 2 \times 11$
   $10 + 2 \times 11$
   $10 + 22$
   32

Page 37 Continued

5) $49 - 14 + 7 \times 5$
   $49 - 14 + 35$
   $35 + 35$
   70
6) $78 + 54 \div 9 \times 7$
   $78 + 6 \times 7$
   $78 + 42$
   120
7) $56 \div 8 - 36 \div 6$
   $7 - 36 \div 6$
   $7 - 6$
   1
8) $5 \times 6 - 18 + 9$
   $30 - 18 + 9$
   $12 + 9$
   21
9) $81 \div 9 + 20 \times 5$
   $9 + 20 \times 5$
   $9 + 100$
   109
10) $9 \times 5 - 15 + 47$
    $45 - 15 + 47$
    $30 + 47$
    77
11) $52 + 63 \div 7 \times 9$
    $52 + 9 \times 9$
    $52 + 81$
    133
12) $168 + 60 \div 12 - 10$
    $168 + 5 - 10$
    $173 - 10$
    163
13) $235 - 14 + 3 \times 5$
    $235 - 14 + 15$
    $221 + 15$
    236
14) $200 \div 10 - 20 + 65$
    $20 - 20 + 65$
    $0 + 65$
    65
15) $235 - 30 + 40 \times 5$
    $235 - 30 + 200$
    $205 + 200$
    405
16) $53 + 80 - 4 \times 3$
    $53 + 80 - 12$
    $133 - 12$
    121
17) $114 + 30 - 36 \div 12$
    $114 + 30 - 3$
    $144 - 3$
    141
18) $72 \div 8 + 11 \times 4$
    $9 + 11 \times 4$
    $9 + 44$
    53
19) $9 \times 6 - 20 + 25$
    $54 - 20 + 25$
    $34 + 25$
    59
20) $99 + 45 \div 9 + 15$
    $99 + 5 + 15$
    $104 + 15$
    119
21) $72 \div 8 - 6 \times 1$
    $9 - 6 \times 1$
    $9 - 6$
    3

Page 38

1) $2 \times 7 - 45 \div 5$
   $14 - 45 \div 5$
   $14 - 9$
   5
2) $19 + 72 \div 6 + 15$
   $19 + 12 + 15$
   $31 + 15$
   46
3) $30 \times 8 - 130 + 60$
   $240 - 130 + 60$
   $110 + 60$
   170
4) $48 \div 6 + 7 \times 4$
   $8 + 7 \times 4$
   $8 + 28$
   36
5) $110 + 43 - 72 \div 8$
   $110 + 43 - 9$
   $153 - 9$
   144
6) $36 \div 4 - 4 \times 2$
   $9 - 4 \times 2$
   $9 - 8$
   1
7) $69 + 127 - 8 \times 7$
   $69 + 127 - 56$
   $196 - 56$
   140
8) $413 - 200 + 5 \times 10$
   $413 - 200 + 50$
   $213 + 50$
   263
9) $9 \times 2 - 72 \div 9$
   $18 - 72 \div 9$
   $18 - 8$
   10
10) $28 + 74 - 5 \times 2$
    $28 + 74 - 10$
    $102 - 10$
    92
11) $61 + 35 - 72 \div 8$
    $61 + 35 - 9$
    $96 - 9$
    87
12) $5 \times 10 - 6 \div 2$
    $50 - 6 \div 2$
    $50 - 3$
    47
13) $12 + 8 \div 4 - 1$
    $12 + 2 - 1$
    $14 - 1$
    13
14) $100 \div 10 + 50 \times 2$
    $10 + 50 \times 2$
    $10 + 100$
    110
15) $80 - 40 \div 4 \div 2$
    $80 - 10 \div 2$
    $80 - 5$
    75
16) $490 - 140 - 12 \times 12$
    $490 - 140 - 144$
    $350 - 144$
    206
17) $25 \div 5 \times 9 - 15$
    $5 \times 9 - 15$
    $45 - 15$
    30
18) $10 \times 6 - 15 \div 5$
    $60 - 15 \div 5$
    $60 - 3$
    57

103) Answer Key

Page 38 Continued

19) $10 + 100 \div 100 - 9$
$10 + 1 - 9$
$11 - 9$
2

20) $201 - 190 + 2 \times 0$
$201 - 190 + 0$
$11 + 0$
11

21) $99 \div 9 - 3 \times 3$
$11 - 3 \times 3$
$11 - 9$
2

22) $11 \times 8 + 24 \div 2$
$88 + 24 \div 2$
$88 + 12$
100

23) $24 - 12 \div 3 \times 4$
$24 - 4 \times 4$
$24 - 16$
8

24) $98 + 72 \div 6 - 10$
$98 + 12 - 10$
$110 - 10$
100

Page 39

1) $12 \times 4 - 54 \div 9$
$48 - 54 \div 9$
$48 - 6$
42

2) $11 \times 7 + 34 - 10$
$77 + 34 - 10$
$111 - 10$
101

3) $69 + 51 - 8 \times 4$
$69 + 51 - 32$
$120 - 32$
88

4) $3 \times 50 - 30 + 20$
$150 - 30 + 20$
$120 + 20$
140

5) $16 + 100 \div 10 - 14$
$16 + 10 - 14$
$26 - 14$
12

6) $35 - 63 \div 9 \times 4$
$35 - 7 \times 4$
$35 - 28$
7

7) $7 \times 2 + 88 \div 11$
$14 + 88 \div 11$
$14 + 8$
22

8) $64 \div 8 - 8 \times 1$
$8 - 8 \times 1$
$8 - 8$
0

9) $50 \div 2 \times 5 - 8$
$25 \times 5 - 8$
$125 - 8$
117

10) $132 \div 12 - 6 \times 0$
$11 - 6 \times 0$
$11 - 0$
11

11) $14 + 36 \div 2 - 9$
$14 + 18 - 9$
$32 - 9$
23

12) $100 - 97 + 8 \times 10$
$100 - 97 + 80$
$3 + 80$
83

13) $12 \times 3 - 32 \div 4$
$36 - 32 \div 4$
$36 - 8$
28

14) $42 \div 6 - 3 \times 1$
$7 - 3 \times 1$
$7 - 3$
4

15) $7 \times 3 + 9 - 11$
$21 + 9 - 11$
$30 - 11$
19

16) $24 + 14 - 9 \times 2$
$24 + 14 - 18$
$38 - 18$
20

17) $20 \times 4 - 27 + 8$
$80 - 27 + 8$
$53 + 8$
61

18) $6 + 144 \div 12 - 12$
$6 + 12 - 12$
$18 - 12$
6

19) $96 - 60 \div 6 \times 4$
$96 - 10 \times 4$
$96 - 40$
56

20) $8 \times 11 + 48 - 12$
$88 + 48 - 12$
$136 - 12$
124

21) $7 \times 8 + 4 \div 4$
$56 + 4 \div 4$
$56 + 1$
57

22) $63 \div 7 - 3 \times 3$
$9 - 3 \times 3$
$9 - 9$
0

23) $19 + 14 \div 14 - 18$
$19 + 1 - 18$
$20 - 18$
2

24) $51 - 19 + 8 \times 4$
$51 - 19 + 32$
$32 + 32$
64

Page 40

1) $6 \times 3 - 40 \div 20$
$18 - 40 \div 20$
$18 - 2$
16

2) $21 + 49 \div 7 + 32$
$21 + 7 + 32$
$28 + 32$
60

3) $9 \times 8 - 23 + 11$
$72 - 23 + 11$
$49 + 11$
60

4) $45 \div 9 + 12 \times 4$
$5 + 12 \times 4$
$5 + 48$
53

5) $64 + 16 - 80 \div 10$
$64 + 16 - 8$
$80 - 8$
72

6) $48 \div 4 - 6 \times 2$
$12 - 6 \times 2$
$12 - 12$
0

7) $157 + 121 - 40 \times 6$
$157 + 121 - 240$
$278 - 240$
38

8) $96 - 80 + 9 \times 5$
$96 - 80 + 45$
$16 + 45$
61

9) $35 \div 5 - 2 \times 0$
$7 - 2 \times 0$
$7 - 0$
7

10) $55 \div 5 - 2 \times 4$
$11 - 2 \times 4$
$11 - 8$
3

11) $98 + 28 \div 7 \times 9$
$98 + 4 \times 9$
$98 + 36$
134

12) $4 \times 9 - 35 \div 5$
$36 - 35 \div 5$
$36 - 7$
29

13) $70 \div 10 - 8 + 19$
$7 - 8 + 19$
$-1 + 19$
18

14) $99 \div 9 + 11 \times 3$
$11 + 11 \times 3$
$11 + 33$
44

15) $48 \div 4 - 24 \div 4$
$12 - 24 \div 4$
$12 - 6$
6

16) $49 - 14 - 6 \times 5$
$49 - 14 - 30$
$35 - 30$
5

17) $70 \div 7 \times 2 - 20$
$10 \times 2 - 20$
$20 - 20$
0

18) $121 \div 11 - 2 \times 4$
$11 - 2 \times 4$
$11 - 8$
3

19) $16 + 88 \div 8 - 9$
$16 + 11 - 9$
$27 - 9$
18

20) $32 - 17 + 4 \times 6$
$32 - 17 + 24$
$15 + 24$
39

21) $84 \div 7 - 6 \times 2$
$12 - 6 \times 2$
$12 - 12$
0

22) $1 \times 13 + 25 \div 5$
$13 + 25 \div 5$
$13 + 5$
18

23) $95 - 15 \div 5 \times 3$
$95 - 3 \times 3$
$95 - 9$
86

24) $33 + 72 \div 6 - 10$
$33 + 12 - 10$
$45 - 10$
35

Page 41

1) $38 - 2 \times 4^2 \div 4$
$38 - 2 \times 16 \div 4$
$38 - 32 \div 4$
$38 - 8$
30

2) $3^2 \times 5 - 2^2 + 6$
$9 \times 5 - 4 + 6$
$45 - 4 + 6$
$41 + 6$
47

3) $9^2 + 9 \times 2 - 18$
$81 + 9 \times 2 - 18$
$81 + 18 - 18$
$99 - 18$
81

4) $2^6 + 2^3 - 8 \div 8$
$64 + 8 - 8 \div 8$
$64 + 8 - 1$
$72 - 1$
71

5) $3^2 + 5^2 \times 4 \div 2$
$9 + 25 \times 4 \div 2$
$9 + 100 \div 2$
$9 + 50$
59

6) $6^2 - 5^2 \times 2 \div 10$
$36 - 25 \times 2 \div 10$
$36 - 50 \div 10$
$36 - 5$
31

7) $13 - 2^2 + 1 \times 2^3$
$13 - 4 + 1 \times 8$
$13 - 4 + 8$
$9 + 8$
17

8) $4^2 \div 2 - 0 \times 2^4$
$16 \div 2 - 0 \times 16$
$8 - 0 \times 16$
$8 - 0$
8

9) $3^2 + 4^2 \div 8 \times 2$
$9 + 16 \div 8 \times 2$
$9 + 2 \times 2$
$9 + 4$
13

10) $5^2 \times 1 + 3^2 - 11$
$25 \times 1 + 9 - 11$
$25 + 9 - 11$
$34 - 11$
23

11) $3^2 \times 2^3 - 55 \div 5$
$9 \times 8 - 55 \div 5$
$72 - 55 \div 5$
$72 - 11$
61

12) $19 - 4^2 \div 8 \times 3$
$19 - 16 \div 8 \times 3$
$19 - 2 \times 3$
$19 - 6$
13

Page 42

1) $3^2 + 5^2 - 5 \times 4$
$9 + 25 - 5 \times 4$
$9 + 25 - 20$
$34 - 20$
14

2) $2^3 \div 2 + 17 \times 1$
$8 \div 2 + 17 \times 1$
$4 + 17 \times 1$
$4 + 17$
21

3) $9^2 \times 0 + 2^4 - 4$
$81 \times 0 + 16 - 4$
$0 + 16 - 4$
$16 - 4$
12

4) $11 + 3^3 \div 3^2 + 7$
$11 + 27 \div 9 + 7$
$11 + 3 + 7$
$14 + 7$
21

5) $8^2 \div 2^3 - 4 \times 2$
$64 \div 8 - 4 \times 2$
$8 - 4 \times 2$
$8 - 8$
0

6) $10^2 \div 4 \times 2 - 2^2$
$100 \div 4 \times 2 - 4$
$25 \times 2 - 4$
$50 - 4$
46

7) $8^2 \div 2 - 1 \times 2^2$
$64 \div 2 - 1 \times 4$
$32 - 1 \times 4$
$32 - 4$
28

8) $2^4 + 5^2 \div 5 - 10$
$16 + 25 \div 5 - 10$
$16 + 5 - 10$
$21 - 10$
11

9) $2^5 - 17 + 2^2 \times 6$
$32 - 17 + 4 \times 6$
$32 - 17 + 24$
$15 + 24$
39

10) $3^4 \div 9 - 3^2 \times 1$
$81 \div 9 - 9 \times 1$
$9 - 9 \times 1$
$9 - 9$
0

11) $3^3 + 12 - 10^2 \div 10$
$27 + 12 - 100 \div 10$
$27 + 12 - 10$
$39 - 10$
29

12) $64 \div 2^3 + 7 \times 2^2$
$64 \div 8 + 7 \times 4$
$8 + 7 \times 4$
$8 + 28$
36

13) $9 \times 6 - 29 + 2^5$
$9 \times 6 - 29 + 32$
$54 - 29 + 32$
$25 + 32$
57

14) $4^2 + 81 \div 3^2 + 15$
$16 + 81 \div 9 + 15$
$16 + 9 + 15$
$25 + 15$
40

15) $88 \div 2^3 - 3 \times 2$
$88 \div 8 - 3 \times 2$
$11 - 3 \times 2$
$11 - 6$
5

Page 43

1) $6^2 + 0 - 2^3 \times 3$
$36 + 0 - 8 \times 3$
$36 + 0 - 24$
$36 - 24$
12

2) $3^3 + 7 - 1^5 \div 1$
$27 + 7 - 1 \div 1$
$27 + 7 - 1$
$34 - 1$
33

3) $7^2 \times 1 - 4^2 + 3$
$49 \times 1 - 16 + 3$
$49 - 16 + 3$
$33 + 3$
36

4) $15 + 8^2 \div 8 + 11$
$15 + 64 \div 8 + 11$
$15 + 8 + 11$
$23 + 11$
34

5) $9^2 \div 9 - 2^2 \times 2$
$81 \div 9 - 4 \times 2$
$9 - 4 \times 2$
$9 - 8$
1

6) $6^2 \div 6 + 14 \times 1$
$36 \div 6 + 14 \times 1$
$6 + 14 \times 1$
$6 + 14$
20

7) $9 \div 3^2 + 6 \times 2^2$
$9 \div 9 + 6 \times 4$
$1 + 6 \times 4$
$1 + 24$
25

8) $7^2 \times 1^{10} - 2^2 + 5$
$49 \times 1 - 4 + 5$
$49 - 4 + 5$
$45 + 5$
50

9) $4^2 + 90 \div 3^2 + 14$
$16 + 90 \div 9 + 14$
$16 + 10 + 14$
$26 + 14$
40

10) $80 \div 2^3 - 3 \times 2$
$80 \div 8 - 3 \times 2$
$10 - 3 \times 2$
$10 - 6$
4

11) $7^2 \div 7 \times 4 - 3^3$
$49 \div 7 \times 4 - 27$
$7 \times 4 - 27$
$28 - 27$
1

12) $2^4 \div 8 - 0 \times 2^2$
$16 \div 8 - 0 \times 4$
$2 - 0 \times 4$
$2 - 0$
2

13) $5^2 + 5^2 \div 5 - 13$
$25 + 25 \div 5 - 13$
$25 + 5 - 13$
$30 - 13$
17

14) $8^2 - 20 + 2^2 \times 10$
$64 - 20 + 4 \times 10$
$64 - 20 + 40$
$44 + 40$
84

15) $4^2 \div 2^3 + 2^2 \times 1$
$16 \div 8 + 4 \times 1$
$2 + 4 \times 1$
$2 + 4$
6

104) Answer Key

Page 44:

1) $39 - 9 \div 1^7 \times 2^2$
$39 - 9 \div 1 \times 4$
$39 - 9 \times 4$
$39 - 36$
3

2) $3^3 \times 8 - 2^4 \div 2$
$27 \times 8 - 16 \div 2$
$216 - 16 \div 2$
$216 - 8$
208

3) $4^2 \div 8 - 2^2 + 39$
$16 \div 8 - 4 + 39$
$2 - 4 + 39$
$-2 + 39$
37

4) $2^4 \times 1^9 \div 8 + 57$
$16 \times 1 \div 8 + 57$
$16 \div 8 + 57$
$2 + 57$
59

5) $9^2 + 9 - 56 \div 2^3$
$81 + 9 - 56 \div 8$
$81 + 9 - 7$
$90 - 7$
83

6) $72 \div 8 - 4 \times 1^3$
$9 - 4 \times 1$
$9 - 4$
5

7) $5^2 \div 5 - 1 \times 2^2$
$25 \div 5 - 1 \times 4$
$5 - 1 \times 4$
$5 - 4$
1

8) $9^2 + 7^2 \div 7 - 12$
$81 + 49 \div 7 - 12$
$81 + 7 - 12$
$88 - 12$
76

9) $2^2 \times 2^3 + 8^2 - 60$
$4 \times 8 + 64 - 60$
$32 + 64 - 60$
$96 - 60$
36

10) $24 \div 2^3 + 2^2 - 1$
$24 \div 8 + 4 - 1$
$3 + 4 - 1$
$7 - 1$
6

11) $6^2 \div 9 + 9 \times 1$
$36 \div 9 + 9 \times 1$
$4 + 9 \times 1$
$4 + 9$
13

12) $9 - 3^2 + 8 \times 2^2$
$9 - 9 + 8 \times 4$
$9 - 9 + 32$
$0 + 32$
32

13) $7^2 + 1^{10} - 2^2 \times 5$
$49 + 1 - 4 \times 5$
$49 + 1 - 20$
$50 - 20$
30

14) $9^2 + 40 \div 2^2 + 9$
$81 + 40 \div 4 + 9$
$81 + 10 + 9$
$91 + 9$
100

15) $25 \div 5^2 - 1^9 \times 1$
$25 \div 25 - 1 \times 1$
$1 - 1 \times 1$
$1 - 1$
0

Page 45:

1) $55 + 6^2 - 3^2 \times 3$
$55 + 36 - 9 \times 3$
$55 + 36 - 27$
$91 - 27$
64

2) $5^2 + 4^2 - 2^2 \div 1$
$25 + 16 - 4 \div 1$
$25 + 16 - 4$
$41 - 4$
37

3) $7^2 - 40 - 6^2 \div 4$
$49 - 40 - 36 \div 4$
$49 - 40 - 9$
$9 - 9$
0

4) $64 + 10^2 \div 10 - 31$
$64 + 100 \div 10 - 31$
$64 + 10 - 31$
$74 - 31$
43

5) $54 \div 3^2 - 1^8 \times 3$
$54 \div 9 - 1 \times 3$
$6 - 1 \times 3$
$6 - 3$
3

6) $63 \div 7 \times 5 - 3^3$
$63 \div 7 \times 5 - 27$
$9 \times 5 - 27$
$45 - 27$
18

7) $0 \div 12 - 8 \times 2^2$
$0 \div 12 - 8 \times 4$
$0 - 8 \times 4$
$0 - 32$
$- 32$

8) $100 + 7^2 \div 7 - 3^3$
$100 + 49 \div 7 - 27$
$100 + 7 - 27$
$107 - 27$
80

9) $65 - 8^2 + 2^2 \times 5$
$65 - 64 + 4 \times 5$
$65 - 64 + 20$
$1 + 20$
21

10) $4^2 + 2^3 - 6 \times 2^2$
$16 + 8 - 6 \times 4$
$16 + 8 - 24$
$24 - 24$
0

11) $9^2 - 8 \times 5 + 1$
$81 - 8 \times 5 + 1$
$81 - 40 + 1$
$41 + 1$
42

12) $90 \div 3^2 + 6 \times 1^{12}$
$90 \div 9 + 6 \times 1$
$10 + 6 \times 1$
$10 + 6$
16

13) $2^4 + 8 - 24 \div 2^2$
$16 + 8 - 24 \div 4$
$16 + 8 - 6$
$24 - 6$
18

14) $8^2 - 49 \div 7^2 + 16$
$64 - 49 \div 49 + 16$
$64 - 1 + 16$
$63 + 16$
79

15) $35 \div 7 - 2^2 \times 1^9$
$35 \div 7 - 4 \times 1$
$5 - 4 \times 1$
$5 - 4$
1

Page 46:

1) $12 - (3 \times 2^2) \div 4$
$12 - (3 \times 4) \div 4$
$12 - 12 \div 4$
$12 - 3$
9

2) $(3^2) (25 - 2^2 \times 6)$
$9 (25 - 4 \times 6)$
$9 (25 - 24)$
$9 (1)$
9

3) $(9^2 \div 3^2 + 2) \times 7$
$(81 \div 9 + 2) \times 7$
$(9 + 2) \times 7$
11×7
77

4) $(64 \div 2^3) \times 9 - 6^2$
$(64 \div 8) \times 9 - 36$
$(8) \times 9 - 36$
$72 - 36$
36

5) $66 \div 6 \times (5^2 - 2^4)$
$66 \div 6 \times (25 - 16)$
$66 \div 6 \times 9$
11×9
99

6) $(36 - 5^2 + 3) \div 7$
$(36 - 25 + 3) \div 7$
$(11 + 3) \div 7$
$14 \div 7$
2

7) $(4 + 5^2 \div 5)^2 + 19$
$(4 + 25 \div 5)^2 + 19$
$(4 + 5)^2 + 19$
$(9)^2 + 19$
$81 + 19$
100

8) $(2^3 \times 6 - 3) \div 3^2$
$(8 \times 6 - 3) \div 9$
$(48 - 3) \div 9$
$(45) \div 9$
5

9) $48 \div (2^2 + 2) (7)$
$48 \div (4 + 2) (7)$
$48 \div (6) (7)$
$8 (7)$
56

10) $44 + (37 - 28) \times 6$
$44 + (9) \times 6$
$44 + 54$
98

11) $91 + (3^2 \times 10 + 11)$
$91 + (9 \times 10 + 11)$
$91 + (90 + 11)$
$91 + (101)$
192

12) $(6^2 - 5^2) (8 + 3)$
$(36 - 25) (8 + 3)$
$(11) (11)$
121

Page 47:

1) $50 + (3^2 \times 7) - 4$
$50 + (9 \times 7) - 4$
$50 + 63 - 4$
$113 - 4$
109

2) $(3^2 - 5) (2^2 \times 3)$
$(9 - 5) (4 \times 3)$
$(4) (12)$
48

3) $(45 \div 3^2 - 2) 5$
$(45 \div 9 - 2) 5$
$(5 - 2) 5$
$(3) 5$
15

4) $28 + (2^3 \times 7) - 6$
$28 + (8 \times 7) - 6$
$28 + 56 - 6$
$84 - 6$
78

5) $(11 - 15 \div 5)^2 + 19$
$(11 - 3)^2 + 19$
$(8)^2 + 19$
$64 + 19$
83

6) $(42 \div 7 + 2) (5 + 1^5)$
$(6 + 2) (5 + 1)$
$(8) (6)$
48

7) $44 - (201 - 196) 7$
$44 - (5) 7$
$44 - 35$
9

8) $75 + (2^2 \times 11 - 14)$
$75 + (4 \times 11 - 14)$
$75 + (44 - 14)$
$75 + 30$
105

9) $(82 - 71) (6 + 3)$
$(82 - 71) (6 + 3)$
$(11) (9)$
99

10) $(8) (36 \div 9 - 2^2)$
$(8) (36 \div 9 - 4)$
$(8) (4 - 4)$
$(8) (0)$
0

11) $(30 - 5^2) (21 \div 7)$
$(30 - 25) (21 \div 7)$
$(30 - 25) (3)$
$(5) (3)$
15

12) $(63 \div 7 - 5)^2 + 86$
$(9 - 5)^2 + 86$
$(4)^2 + 86$
$16 + 86$
102

13) $(8 \times 6 - 21) \div 3^2$
$(8 \times 6 - 21) \div 9$
$(48 - 21) \div 9$
$27 \div 9$
3

14) $43 + 4 (83 - 78)$
$43 + 4 (5)$
$43 + 20$
63

15) $(27 \div 9 + 2) (3^2 + 1^6)$
$(27 \div 9 + 2) (9 + 1)$
$(3 + 2) (9 + 1)$
$(5) (10)$
50

Page 48:

1) $(15 + 6 \bullet 8) \div 7$
$(15 + 48) \div 7$
$63 \div 7$
9

2) $(3^2 \bullet 11 - 74) + 25$
$(9 \bullet 11 - 74) + 25$
$(99 - 74) + 25$
$25 + 25$
50

3) $(4^2 - 6) (5 \bullet 2)$
$(16 - 6) (5 \bullet 2)$
$(16 - 6) (10)$
$(10) (10)$
100

4) $77 + (85 - 79) \bullet 6$
$77 + 6 \bullet 6$
$77 + 36$
113

5) $9^2 + (3^2 \bullet 16 \div 8)$
$81 + (9 \bullet 16 \div 8)$
$81 + (9 \bullet 2)$
$81 + 18$
99

6) $134 - (3^2 \bullet 18 \div 9)$
$134 - (9 \bullet 18 \div 9)$
$134 - (9 \bullet 2)$
$134 - 18$
116

7) $(65 - 8^2 + 1) \bullet 2^3$
$(65 - 64 + 1) \bullet 8$
$(1 + 1) \bullet 8$
$2 \bullet 8$
16

8) $(7^2 \bullet 1^{10} - 3^2) - 40$
$(49 \bullet 1 - 9) - 40$
$(49 - 9) - 40$
$40 - 40$
0

9) $6^2 + (15 \div 5) \bullet 7$
$36 + (15 \div 5) \bullet 7$
$36 + 3 \bullet 7$
$36 + 21$
57

10) $(10^2 \bullet 1 + 3^2) - 54$
$(100 \bullet 1 + 9) - 54$
$(100 + 9) - 54$
$109 - 54$
55

11) $3^2 \bullet (2^3 - 66 \div 11)$
$9 \bullet (8 - 66 \div 11)$
$9 \bullet (8 - 6)$
$9 \bullet 2$
18

12) $(22 - 4^2) \div (2 \bullet 3)$
$(22 - 16) \div (2 \bullet 3)$
$(22 - 16) \div 6$
$6 \div 6$
1

Page 49

1) $(55 - 2 \bullet 5^2) 4$
$(55 - 2 \bullet 25) 4$
$(55 - 50) 4$
$(5) 4$
20

2) $3^2 (11 - 3^2 + 4)$
$9 (11 - 9 + 4)$
$9 (2 + 4)$
$9 (6)$
54

3) $(6^2 + 4 \bullet 2) - 28$
$(36 + 4 \bullet 2) - 28$
$(36 + 8) - 28$
$44 - 28$
16

4) $2^5 (2^3 - 8) \div 16$
$32 (8 - 8) \div 16$
$32 (0) \div 16$
$0 \div 16$
0

5) $(1^5 + 2^2) (24 \div 3)$
$(1 + 4) (24 \div 3)$
$(1 + 4) (8)$
$(5) (8)$
40

6) $250 - (5^2 \bullet 2 + 100)$
$250 - (25 \bullet 2 + 100)$
$250 - (50 + 100)$
$250 - 150$
100

7) $(6 + 3^2 - 7) \bullet 2^3$
$(6 + 9 - 7) \bullet 8$
$(15 - 7) \bullet 8$
$8 \bullet 8$
64

8) $50 (2 - 0 \bullet 5^3)$
$50 (2 - 0 \bullet 125)$
$50 (2 - 0)$
$50 (2)$
100

9) $(2^2 + 4^2 \div 16) 7$
$(4 + 16 \div 16) 7$
$(4 + 1) 7$
$(5) 7$
35

10) $9 (1 + 8^2 - 54)$
$9 (1 + 64 - 54)$
$9 (65 - 54)$
$9 (11)$
99

11) $(2^3 \bullet 3^2) - (90 \div 10)$
$(8 \bullet 9) - (90 \div 10)$
$(72) - (90 \div 10)$
$72 - 9$
63

12) $158 - (7^2 \div 49) 8$
$158 - (49 \div 49) 8$
$158 - (1) 8$
$158 - 8$
150

105) Answer Key

Page 50:

1) $(5^2 + 10 - 3^2)\, 2$
$(25 + 10 - 9)\, 2$
$(35 - 9)\, 2$
$(26)\, 2$
52

2) $(7^2 + 31) - (1^5 \cdot 24)$
$(49 + 31) - (1 \cdot 24)$
$(49 + 31) - (24)$
$80 - 24$
56

3) $6\, (25 - 4^2 - 5)$
$6\, (25 - 16 - 5)$
$6\, (9 - 5)$
$6\, (4)$
24

4) $(70 - 8^2)\, (32 \div 4)$
$(70 - 64)\, (32 \div 4)$
$(70 - 64)\, (8)$
$(6)\, (8)$
48

5) $44 \div (47 - 6^2) \cdot 2$
$44 \div (47 - 36) \cdot 2$
$44 \div (11) \cdot 2$
$4 \cdot 2$
8

6) $9^2 + (8 \cdot 4 - 26)$
$81 + (8 \cdot 4 - 26)$
$81 + (32 - 26)$
$81 + 6$
87

7) $(4^2 \div 2 - 1) \cdot 2^2$
$(16 \div 2 - 1) \cdot 4$
$(8 - 1) \cdot 4$
$(7) \cdot 4$
28

8) $82 + (60 \div 5) - 33$
$82 + 12 - 33$
$94 - 33$
61

9) $(5^2 - 20 + 2^2)\, 10$
$(25 - 20 + 4)\, 10$
$(5 + 4)\, 10$
$(9)\, 10$
90

10) $(3^2 \cdot 11) - (57 \times 1)$
$(9 \cdot 11) - (57 \times 1)$
$99 - 57$
42

11) $54 + (6 \cdot 5)\, 3$
$54 + (30)\, 3$
$54 + 90$
144

12) $9\, (34 - 7 \cdot 2^2)$
$9\, (34 - 7 \cdot 4)$
$9\, (34 - 28)$
$9\, (6)$
54

13) $(9^2 \cdot 1^{10} - 7^2) + 105$
$(81 \cdot 1 - 49) + 105$
$(81 - 49) + 105$
$32 + 105$
137

14) $4^2 + (20 \div 2^2 + 14)$
$16 + (20 \div 4 + 14)$
$16 + (5 + 14)$
$16 + 19$
35

15) $(40 \div 2^2) + (9 \times 7)$
$(40 \div 4) + (9 \times 7)$
$10 + (9 \times 7)$
$10 + 63$
73

Page 51

1) $8 - (2 \cdot (16 \div 4))$
$8 - (2 \cdot 4)$
$8 - 8$
0

2) $(88 - (10 - 2)) \div 8$
$(88 - 8) \div 8$
$80 \div 8$
10

3) $(4 + (10 - (3 + 5)))\, 7$
$(4 + (10 - 8))\, 7$
$(4 + 2)\, 7$
$(6)\, 7$
42

4) $70 + (2\, (9 \div 3 + 2)^2)$
$70 + (2\, (3 + 2)^2)$
$70 + (2\, (5)^2)$
$70 + 2\, (25)$
$70 + 50$
120

5) $183 + (3\, (7 - 4)^2)$
$183 + (3\, (3)^2)$
$183 + 3\, (9)$
$183 + 27$
210

6) $5\, (3^2 \div 3)^2 + 6)$
$5\, ((9 \div 3)^2 + 6)$
$5\, (3)^2 + 6)$
$5\, (9 + 6)$
$5\, (15)$
75

7) $87 - ((24 \div 8)^2 + 31)$
$87 - ((3)^2 + 31))$
$87 - (9 + 31)$
$87 - 40$
47

8) $(9 + (16 + 10))\, 5$
$(9 + 26)\, 5$
$(35)\, 5$
175

9) $((1^{10} + 3^2) \div 5)\, 7$
$((1 + 9) \div 5)\, 7$
$(10 \div 5)\, 7$
$(2)\, 7$
14

10) $(12 - 3^2)\, 2)^2 - 31$
$((12 - 9)\, 2)^2 - 31$
$((3)\, 2)^2 - 31$
$(6)^2 - 31$
$36 - 31$
5

11) $3\, (8 + 2) \cdot 5 - 45)$
$3\, (10 \cdot 5 - 45)$
$3\, (50 - 45)$
$3\, (5)$
15

12) $18 - (1 + (3^3 - 22))$
$18 - (1 + (27 - 22))$
$18 - (1 + 5)$
$18 - 6$
12

Page 52:

1) $(92 - ((14 - 8) \div 2)\, 3)$
$(92 - (6 \div 2)\, 3)$
$(92 - (3)\, 3)$
$92 - 9$
83

2) $(4\, (3 + 7) - 6^2) \div 2$
$(4\, (3 + 7) - 36) \div 2$
$(4\, (10) - 36) \div 2$
$(40 - 36) \div 2$
$4 \div 2$
2

3) $((7^2 \cdot 1) - 35) + 8\, (6)$
$((49 \cdot 1) - 35) + 8\, (6)$
$(49 - 35) + 48$
$14 + 48$
62

4) $(34 - (10^2 \div 100)) + 9$
$(34 - (100 \div 100)) + 9$
$(34 - 1) + 9$
$33 + 9$
42

5) $((8 \cdot 9) \div 8)^2 - 31$
$(72 \div 8)^2 - 31$
$(9)^2 - 31$
$81 - 31$
50

6) $111 + (30\, (36 \div 9))$
$111 + (30\, (4))$
$111 + 120$
231

7) $(56 \div (2^2 + 3)) \cdot 4$
$(56 \div (4 + 3)) \cdot 4$
$(56 \div 7) \cdot 4$
$(8) \cdot 4$
32

8) $(2\, (103 - 96)) + 46$
$(2\, (7)) + 46$
$14 + 46$
60

9) $(1^{10} + (40 \div 5)) \cdot 6$
$(1 + 8) \cdot 6$
$(9) \cdot 6$
54

10) $20 \div (2^3 - (2^2 \cdot 1))$
$20 \div (8 - (4 \cdot 1))$
$20 \div (8 - 4)$
$20 \div 4$
5

11) $(35 \div (21 - 14))\, 6$
$(35 \div 7)\, 6$
$(5)\, 6$
30

12) $((16 \div 2^4) + 4) - 5$
$((16 \div 16) + 4) - 5$
$(1 + 4) - 5$
$5 - 5$
0

13) $13 \cdot (21 - (4^2 + 5))$
$13 \cdot (21 - (16 + 5))$
$13 \cdot (21 - 21)$
$13 \cdot 0$
0

14) $16 + (8\, (25 \div 5^2) + 9)$
$16 + (8\, (25 \div 25) + 9)$
$16 + (8\, (1) + 9)$
$16 + (8 + 9)$
$16 + 17$
33

15) $(80 \div (57 - 49)) \cdot 6$
$(80 \div 8) \cdot 6$
$(10) \cdot 6$
60

Page 53:

1) $((54 - 29) - 2^2) \div 7$
$(25 - 2^2) \div 7$
$(25 - 4) \div 7$
$21 \div 7$
3

2) $(5 + (16 - (4 + 3^2))\, 7$
$(5 + (16 - (4 + 9))\, 7$
$(5 + (16 - 13))\, 7$
$(5 + 3)\, 7$
$(8)\, 7$
56

3) $45 \div (11 - 2^3)\, 3)$
$45 \div (11 - 8)\, 3)$
$45 \div (3)\, 3$
$45 \div 9$
5

4) $105 - (4\, (19 - 11))$
$105 - (4\, (8))$
$105 - 32$
73

5) $26 + ((15 - 9)\, 2 - 12)$
$26 + ((6)\, 2 - 12)$
$26 + (12 - 12)$
$26 + 0$
26

6) $(49 - 7\, (31 - 5^2)) \cdot 9$
$(49 - 7\, (31 - 25)) \cdot 9$
$(49 - 7\, (6)) \cdot 9$
$(49 - 42) \cdot 9$
$7 \cdot 9$
63

7) $(35 \div (8 - 1) \cdot 2^2$
$(35 \div (8 - 1)) \cdot 4$
$(35 \div 7) \cdot 4$
$5 \cdot 4$
20

8) $((25 + 5^2) \div 10)^2 - 19$
$((25 + 25) \div 10)^2 - 19$
$(50 \div 10)^2 - 19$
$(5)^2 - 19$
$25 - 19$
6

9) $(8(6) - (12 + 6^2)) \cdot 13$
$(48 - (12 + 36)) \cdot 13$
$(48 - 48) \cdot 13$
$(0) \cdot 13$
0

10) $((28 \div 4)\, (36 - 9)) + 2$
$(7)\, (36 - 9)) + 2$
$7(27) + 2$
$189 + 2$
191

11) $(40 \div (7 + 13))\, 9$
$(40 \div 20)\, 9$
$(2)\, 9$
18

12) $49 \div (3^2 + (10 \times 2^2))$
$49 \div (9 + (10 \times 4))$
$49 \div (9 + 40)$
$49 \div 49$
1

13) $(8\, (4^2 - 2^2)) + 3(7)$
$(8\, (16 - 4)) + 3(7)$
$(8\, (16 - 4)) + 21$
$8\, (12) + 21$
$96 + 21$
117

14) $((5 + 4)\, (16 - 9)) \div 7$
$(9)\, (16 - 9)) \div 7$
$(9\, (7)) \div 7$
$63 \div 7$
9

15) $6\, (2\, (19 - 2^3) - (4 \cdot 3))$
$6\, (2\, (19 - 8) - (4 \cdot 3))$
$6\, (2\, (19 - 8) - 12)$
$6\, (2\, (11) - 12)$
$6\, (22 - 12)$
$6\, (10)$
60

Page 54:

1) $(3^2 \cdot (28 \div 7)) - 5(4)$
$(9 \cdot (28 \div 7)) - 5(4)$
$(9 \cdot 4) - 5(4)$
$36 - 20$
16

2) $(9^2\, (10 - 9) - 54) + 67$
$(81\, (10 - 9) - 54) + 67$
$(81\, (1) - 54) + 67$
$(81 - 54) + 67$
$27 + 67$
94

3) $112 + (4^2 - 2(25 \div 5))$
$112 + (16 - 2(25 \div 5))$
$112 + (16 - 2(5))$
$112 + (16 - 10)$
$112 + 6$
118

4) $(30 - 16) \div 2) \cdot 3$
$(14 \div 2) \cdot 3$
$7 \cdot 3$
21

5) $100 \div ((29 - 24)\, 2)$
$100 \div (5)\, 2$
$100 \div 10$
10

6) $59 - ((63 - 6^2) + 15)$
$59 - ((63 - 36) + 15)$
$59 - (27 + 15)$
$59 - 42$
17

7) $(7\, (32 \div 2^3) - 25) \cdot 2^2$
$(7\, (32 \div 8) - 25) \cdot 4$
$(7\, (4) - 25) \cdot 4$
$(28 - 25) \cdot 4$
$3 \cdot 4$
12

8) $165 + (9(21 \div 7) - 17)$
$165 + (9(3) - 17)$
$165 + (27 - 17)$
$165 + (10)$
175

9) $(22 - (13 + 2^2))\, 9$
$(22 - (13 + 4)\, 9$
$(22 - 17)\, 9$
$(5)\, 9$
45

10) $(7(16 \div 2^3) - 2^2)\, 5$
$(7(16 \div 8) - 4)\, 5$
$(7(2) - 4)\, 5$
$(14 - 4)\, 5$
$(10)\, 5$
50

11) $(5\, (42 \div 6)) - 8(4 + 2)$
$(5\, (7)) - 8(4 + 2)$
$(5\, (7)) - 8(6)$
$35 - 48$
-13

12) $32 \div ((48 \div 6)\, 2^2)$
$32 \div ((48 \div 6)\, 4)$
$32 \div ((8)\, 4)$
$32 \div 32$
1

13) $((68 \times 1^{10}) - 47) + 23$
$((68 \times 1) - 47) + 23$
$(68 - 47) + 23$
$21 + 23$
44

14) $((29 + 11) \div 2^2) + 89$
$((29 + 11) \div 4) + 89$
$((40) \div 4) + 89$
$10 + 89$
99

15) $((54 \div 3^2) - 3)\, 6$
$((54 \div 9) - 3)\, 6$
$(6 - 3)\, 6$
$(3)\, 6$
18

Page 55:

1) $((2^3 \times 6) \div (3 + 5))\, 9$
$((8 \times 6) \div (3 + 5))\, 9$
$((8 \times 6) \div 8)\, 9$
$(48 \div 8)\, 9$
$(6)\, 9$
54

2) $(9 + (3 + 6)^2) - 37$
$(9 + (9)^2) - 37$
$(9 + 81) - 37$
$90 - 37$
53

3) $314 - (3\, (2 \times 3) + 25)$
$314 - (3\, (6) + 25)$
$314 - (18 + 25)$
$314 - 43$
271

4) $(2 + (10^2 - 94^2)) + 40$
$(2 + (100 - 94^2)) + 40$
$(2 + (6)^2) + 40$
$(2 + 36) + 40$
$38 + 40$
78

5) $26 + ((64 \div 8) - 4)^2$
$26 + (8 - 4)^2$
$26 + (4)^2$
$26 + 16$
42

6) $(2 + (54 \div 3^2)) \cdot 7$
$(2 + (54 \div 9)) \cdot 7$
$(2 + 6) \cdot 7$
$8 \cdot 7$
56

7) $6^2 + ((1 + 7^2) - 11)$
$36 + ((1 + 49) - 11)$
$36 + (50 - 11)$
$36 + 39$
75

8) $(2^3 + (92 - 60)) \div 5$
$(8 + (92 - 60)) \div 5$
$(8 + 32) \div 5$
$40 \div 5$
8

9) $((25 - 4^2)\, 2) + 3(6 + 4)$
$(25 - 16)\, 2) + 3(6 + 4)$
$(25 - 16)\, 2) + 3(10)$
$(9)2) + 3(10)$
$18 + 30$
48

106) Answer Key

Page 55 Continued:

10) $102 - (47 - (36 \div 6))$
$102 - (47 - 6)$
$102 - 41$
61

12) $(6 + (66 \div 11)) - 12$
$(6 + 6) - 12$
$12 - 12$
0

14) $(8(16 - 7) + 48) - 12$
$(8(9) + 48) - 12$
$(72 + 48) - 12$
$120 - 12$
108

11) $((30 \cdot 2) - 27) + 2^3$
$((30 \cdot 2) - 27) + 8$
$(60 - 27) + 8$
$33 + 8$
41

13) $((108 - 60) \div 6) \div 2^2$
$((108 - 60) \div 6) \div 4$
$(48 \div 6) \div 4$
$8 \div 4$
2

15) $7(2^3) + (5(2)(28 \div 4))$
$7(8) + (5(2)(28 \div 4))$
$7(8) + (5(2)(7))$
$56 + 10(7)$
$56 + 70$
126

Page 56:

1) $19 + ((8 - 2^3) \times 3)$
$19 + ((8 - 8) \times 3)$
$19 + (0 \times 3)$
$19 + 0$
19

6) $((6(3) + 45) \div 63) + 1$
$((18 + 45) \div 63) + 1$
$(63 \div 63) + 1$
$1 + 1$
2

11) $9(7) \div (25 - (2 \times 9))$
$63 \div (25 - (2 \times 9))$
$63 \div (25 - 18)$
$63 \div 7$
9

2) $3^3 + (35 \div (1^5 \times 7))$
$27 + (35 \div (1 \times 7))$
$27 + (35 \div 7)$
$27 + 5$
32

7) $((12 \div 4) 9) \div 3) + 12$
$(((3)9) \div 3) + 12$
$(27 \div 3) + 12$
$9 + 12$
21

12) $((34 - 5^2)^2 + 40) - 99$
$((34 - 25)^2 + 40) - 99$
$((9)^2 + 40) - 99$
$(81 + 40) - 99$
$121 - 99$
22

3) $((164 - 85) - 4^2) + 39$
$((164 - 85) - 16) + 39$
$(79 - 16) + 39$
$63 + 39$
102

8) $7(42 \div 6) - 3) + 1^{10}$
$(7(42 \div 6) - 3) + 1$
$(7(7) - 3) + 1$
$(49 - 3) + 1$
$46 + 1$
47

13) $(21 \div 3) 6 + 26) - 54$
$((7)6 + 26) - 54$
$(42 + 26) - 54$
$68 - 54$
14

4) $(95 + (20 \div 4)) - 3$
$(95 + 5) - 3$
$100 - 3$
97

9) $7((3 + 10) - 11) \div 2)$
$7((13 - 11) \div 2)$
$7(2 \div 2)$
$7(1)$
7

14) $(26 + 34) \div 6) + 54$
$(60 \div 6) + 54$
$10 + 54$
64

5) $79 - ((24 \div 4) 4 + 12)$
$79 - ((6) 4 + 12)$
$79 - (24 + 12)$
$79 - (36)$
43

10) $(31 + (18 - 4^2) 5) - 22$
$(31 + (18 - 16) 5) - 22$
$(31 + (2) 5) - 22$
$(31 + 10) - 22$
$41 - 22$
19

15) $(56 \div 2^3) - 4) 2 - 5$
$(56 \div 8) - 4) 2 - 5$
$(7 - 4) 2 - 5$
$(3) 2 - 5$
$6 - 5$
1

Page 57:

1) $(8(11 - 3^2)) + 5(4 \cdot 9)$
$(8(11 - 9)) + 5(4 \cdot 9)$
$(8(11 - 9)) + 5(36)$
$(8(2)) + 5(36)$
$16 + 180$
196

6) $(19 + (1 + 7)^2) - 40$
$(19 + (8)^2) - 40$
$(19 + 64) - 40$
$83 - 40$
43

11) $(5 + (8 \div 2^2)) \cdot 9$
$(5 + (8 \div 4)) \cdot 9$
$(5 + 2) \cdot 9$
$7 \cdot 9$
63

2) $((122 - 67) + 26) \div 9$
$(55 + 26) \div 9$
$81 \div 9$
9

7) $500 - (5(4 \times 2) + 125)$
$500 - (5(8) + 125)$
$500 - (40 + 125)$
$500 - 165$
335

12) $((46 + 4) - 7^2) - 1^8 + 3$
$((46 + 4) - 49) - 1 + 3$
$(50 - 49) - 1 + 3$
$1 - 1 + 3$
$0 + 3$
3

3) $(10 + (24 \div 8) - 3) 6$
$(10 + 3 - 3) 6$
$(13 - 3) 6$
$(10) 6$
60

8) $(11 + (6^2 - 31)^2) - 35$
$(11 + (36 - 31)^2) - 35$
$(11 + (5)^2) - 35$
$(11 + 25) - 35$
$36 - 35$
1

13) $3^3 + ((92 - 52) \div 4)$
$27 + ((92 - 52) \div 4)$
$27 + (40 \div 4)$
$27 + (10)$
37

4) $(10(4 + 2) - 12) \div 2^3$
$(10(4 + 2) - 12) \div 8$
$(10(6) - 12) \div 8$
$(60 - 12) \div 8$
$48 \div 8$
6

9) $26 + ((42 \div 6) - 5)^2$
$26 + (7 - 5)^2$
$26 + (2)^2$
$26 + 4$
30

14) $(2(35 - 26) + 3) - 2^3$
$(2(35 - 26) + 3) - 8$
$(2(9) + 3) - 8$
$(18 + 3) - 8$
$21 - 8$
13

5) $(51 + 4(23 - 4^2)) - 62$
$(51 + 4(23 - 16)) - 62$
$51 + 4(7) - 62$
$51 + 28 - 62$
$79 - 62$
17

10) $88 - (4(55 \div 11) + 22)$
$88 - (4(5) + 22)$
$88 - (20 + 22)$
$88 - 42$
46

15) $112 - (16 - (14 \div 7))$
$112 - (16 - 2)$
$112 - 14$
98

Page 58:

1) $-5 + -3 \times -6 \div 2$
$-5 + 18 \div 2$
$-5 + 9$
4

5) $(-2)^2 + (-5)^2 - 18 \div -2$
$4 + (-5)^2 - 18 \div -2$
$4 + 25 - 18 \div -2$
$4 + 25 - (-9)$
$29 - (-9)$
38

9) $-(4^2 + 3^2) \div 5 + -7$
$-(16 + 9) \div 5 + -7$
$-25 \div 5 + -7$
$-5 + -7$
-12

2) $(-3)^2 \times 5 - (-2)^2 + 6$
$9 \times 5 - (-2)^2 + 6$
$9 \times 5 - 4 + 6$
$45 - 4 + 6$
$41 + 6$
47

6) $6^2 + 3^2 \times -2 \div 9$
$36 + 9 \times -2 \div 9$
$36 + (-18) \div 9$
$36 + -2$
34

10) $12 - (-9) - 3^2 + 11$
$12 - (-9) - 9 + 11$
$12 + 9 - 9 + 11$
$21 - 9 + 11$
$12 + 11$
23

3) $2^2 + -9 \times 2 - (-10)$
$4 + -9 \times 2 - (-10)$
$4 + -18 + 10$
$-14 + 10$
-4

7) $-13 + 2^2 + 1 \times (-2)^3$
$-13 + 4 + 1 \times -8$
$-13 + 4 + (-8)$
$-9 + -8$
-17

11) $9 + (8 \div -2 - 3)(-5)$
$9 + (-4 - 3)(-5)$
$9 + (-7)(-5)$
$9 + 35$
44

4) $-4 + (-2)^5 - 12 \div -3$
$-4 + -32 - 12 \div -3$
$-4 + -32 + 4$
$-36 + 4$
-32

8) $(-4)^2 \div 8 - (-5) \times 3$
$16 \div 8 - (-5) \times 3$
$16 \div 8 - (-15)$
$2 - (-15)$
$2 + 15$
17

12) $7(24 \div -6 \times 2) - (-56)$
$7(-4 \times 2) - (-56)$
$7(-8) - (-56)$
$-56 + 56$
0

Page 59

1) $(7^2 - 45 + 2^2)(-6)$
$(49 - 45 + 4)(-6)$
$(4 + 4)(-6)$
$(8)(-6)$
-48

6) $6^2 - 7 \cdot 3 + -26$
$36 - 21 + -26$
$15 + -26$
-11

11) $24 - (7 - 9)(-3)$
$24 - (-2)(-3)$
$24 - 6$
18

2) $(-8)^2 + 55 + 1^5 \cdot -25$
$64 + 55 + 1 \cdot -25$
$64 + 55 + -25$
$119 + -25$
94

7) $((-2)^3 \div 2 - 6) \cdot 2$
$(-8 \div 2 - 6) \cdot 2$
$(-4 - 6) \cdot 2$
$(-10) \cdot 2$
-20

12) $(9 - 34 + 17) \cdot 2^2$
$(9 - 34 + 17) \cdot 4$
$(-25 + 17) \cdot 4$
$(-8) \cdot 4$
-32

3) $-6(25 - 4^2 - 4) + -22$
$-6(25 - 16 - 4) + -22$
$-6(9 - 4) + -22$
$-6(5) + -22$
$-30 + -22$
-52

8) $58 + -(12 \div 2) - 33$
$58 + -6 - 33$
$52 - 33$
19

13) $(-2)^3 - 1^6 + 7^2 - 15$
$-8 - 1 + 49 - 15$
$-9 + 49 - 15$
$40 - 15$
25

4) $-(70 - 8^2)(-36 \div 4)$
$-(70 - 64)(-36 \div 4)$
$-(6)(-9)$
54

9) $3^2 - 29 + 2^3 - (-17)$
$9 - 29 + 8 - (-17)$
$-20 + 8 + 17$
$-12 + 17$
5

14) $4^2 - (-28 \div 7 + 14)$
$16 - (-28 \div 7 + 14)$
$16 - (-4 + 14)$
$16 - 10$
6

5) $-45 \div (41 - 6^2) \cdot (-2)$
$-45 \div (41 - 36) \cdot (-2)$
$-45 \div (5) \cdot (-2)$
$-9 \cdot (-2)$
18

10) $(-3)^2 \cdot 11 - 77 \cdot -1$
$9 \cdot 11 - 77 \cdot -1$
$99 - 77 \cdot -1$
$99 - (-77)$
$99 + 77$
176

15) $(35 \div -7) - (2 \times 9)$
$-5 - (2 \times 9)$
$-5 - (18)$
-23

Page 60

1) $-81 \div -9 - (-2) \cdot 3$
$9 - (-2) \cdot 3$
$9 - (-6)$
$9 + 6$
15

6) $-24 - 12 \div 3 \times -4$
$-24 - 4 \times -4$
$-24 + 16$
-8

11) $-(-5)^2 \div -5 \cdot 3^2 - 15$
$-25 \div -5 \cdot 9 - 15$
$5 \cdot 9 - 15$
$45 - 15$
30

2) $29 - 63 \div 7 - (-15)$
$29 - 9 - (-15)$
$29 - 9 + 15$
$20 + 15$
35

7) $112 + (-120) - 7 \cdot -8$
$112 + (-120) + 56$
$-8 + 56$
48

12) $-69 + 120 - 2^3 \cdot 7$
$-69 + 120 - 8 \cdot 7$
$-69 + 120 - 56$
$51 - 56$
-5

3) $3 \cdot -8 - 13 + 37 - 11$
$-24 - 13 + 37 - 11$
$-37 + 37 - 11$
$0 - 11$
-11

8) $-10 \div 10 - (-3) \times 11$
$-1 - (-3) \times 11$
$-1 - (-33)$
$-1 + 33$
32

13) $(-2)^3 \cdot (-1)^5 + 6^2 - (-30)$
$-8 \cdot (-1)^5 + 6^2 - (-30)$
$-8 \cdot -1 + 36 - (-30)$
$8 + 36 - (-30)$
$44 - (-30)$
74

4) $48 \div 8 + -7 \cdot 4 + -6$
$6 + -7 \cdot 4 + -6$
$6 + -28 + -6$
$-22 + -6$
-28

9) $-57 + 42 \div 6 \cdot 7$
$-57 + 7 \cdot 7$
$-57 + 49$
-8

14) $54 \div 6 - 4 \cdot -5$
$9 - 4 \cdot -5$
$9 + 20$
29

5) $105 + -66 - 72 \div 9$
$105 + -66 - 8$
$39 - 8$
31

10) $(-4)^2 + 29 - 38 \cdot 1$
$16 + 29 - 38$
$45 - 38$
7

15) $10^2 - (-28) \div 7 - 104$
$100 - (-28) \div 7 - 104$
$100 - (-4) - 104$
$104 - 104$
0

107) Answer Key

Page 61

1) $41 + 56 \div -7 + (-32)$
$41 + -8 + (-32)$
$33 + (-32)$
1

2) $-80 \div 8 \cdot -2 - 20$
$-10 \cdot -2 - 20$
$20 - 20$
0

3) $-(-4)^2 + 88 \div 8 + (-9)$
$-16 + 88 \div 8 + (-9)$
$-16 + 11 + (-9)$
$-5 + -9$
-14

4) $-9(-1) + -6 + 45 \div -5$
$9 + -6 + 45 \div -5$
$9 + -6 + -9$
$3 + -9$
-6

5) $17 + -16 - 50 \div -5$
$17 + -16 - (-10)$
$1 + 10$
11

6) $-79 + 21 \div -7 \cdot -5$
$-79 + -3 \cdot -5$
$-79 + 15$
-64

7) $250 - 125 + -60 \cdot 3$
$250 - 125 + -180$
$125 + -180$
-55

8) $-32 \div 4 - 2 \times 0 + 4$
$-8 - 2 \times 0 + 4$
$-8 - 0 + 4$
$-8 + 4$
-4

9) $-39 - 11 + -4 \cdot 6$
$-39 - 11 + -24$
$-50 + -24$
-74

10) $-12 - 1 \cdot 9 + (-4)^2 + 37$
$-12 - 1 \cdot 9 + 16 + 37$
$-12 - 9 + 16 + 37$
$-21 + 16 + 37$
$-5 + 37$
32

11) $-(-6)^2 - 5 + 12 - 4$
$-36 - 5 + 12 - 4$
$-41 + 12 - 4$
$-29 - 4$
-33

12) $-33 + 85 - 9 \cdot (-2)^3$
$-33 + 85 - 9 \cdot -8$
$-33 + 85 - (-72)$
$52 - (-72)$
124

13) $(-2)^4 - 22 - (-7) \cdot 2$
$16 - 22 - (-7) \cdot 2$
$16 - 22 - (-14)$
$-6 - (-14)$
8

14) $27 \div -3 - (-5) - 96$
$-9 - (-5) - 96$
$-4 - 96$
-100

15) $-31 - (-26) - (-8) \div 4$
$-31 - (-26) - (-2)$
$-5 - (-2)$
-3

Page 62

1) $-16 \div -2 + (-4) \cdot 9$
$8 + (-4) \cdot 9$
$8 + (-36)$
-28

2) $-127 + 78 - 6 \div 2 - 8$
$-127 + 78 - 3 - 8$
$-49 - 3 - 8$
$-52 - 8$
-60

3) $77 - 113 - (-24) - 13$
$-36 - (-24) - 13$
$-12 - 13$
-25

4) $4^2 - 17 - 36 \div (-2)^2$
$16 - 17 - 36 \div 4$
$16 - 17 - 9$
$-1 - 9$
-10

5) $15 + -30 - (-18) \div -9$
$15 + -30 - 2$
$-15 - 2$
-17

6) $(-9 - 5 + 17 - 2)^2 (-6)$
$(-14 + 15)^2 (-6)$
$(1)^2 (-6)$
-6

7) $314 + (-520) - 9 \cdot 2^3$
$314 + (-520) - 9 \cdot 8$
$314 + (-520) - 72$
$-206 - 72$
-278

8) $- (-9) \times 11 - 10^2 \div 10$
$- (-9) \times 11 - 100 \div 10$
$- (-99) - 100 \div 10$
$99 - 10$
89

9) $-9 (-1 + 8^2 - 54) - 7$
$-9 (-1 + 64 - 54) - 7$
$-9 (63 - 54) - 7$
$-9 (9) - 7$
$-81 - 7$
-88

10) $-14 - 3(2 - 3) + 27$
$-14 - 3(-1) + 27$
$-14 + 3 + 27$
$-11 + 27$
16

11) $-3 \cdot 2^3 + 20 - 82$
$-3 \cdot 8 + 20 - 82$
$-24 + 20 - 82$
$-4 - 82$
-86

12) $-41 - 3^3 - (-7) + (-6)$
$-41 - 27 - (-7) + (-6)$
$-68 - (-7) + (-6)$
$-61 + (-6)$
-67

13) $(-3)^3 + -1^5 + 7^2 - 40$
$-27 + -1 + 49 - 40$
$-28 + 49 - 40$
$21 - 40$
-19

14) $(5^2 - 23 + 2)^2 - (-13)$
$(25 - 23 + 2)^2 - (-13)$
$(2 + 2)^2 - (-13)$
$(4)^2 - (-13)$
$16 - (-13)$
29

15) $75 - (-9)^2 + 7 - (-23)$
$75 - 81 + 7 - (-23)$
$-6 + 7 - (-23)$
$1 - (-23)$
24

Page 63

1) $15 + 2 (6 + ((40 \div 8) - 3) \, 4 - 2)$
$15 + 2 (6 + (5 - 3) \, 4 - 2)$
$15 + 2 (6 + (2) \, 4 - 2)$
$15 + 2 (6 + 8 - 2)$
$15 + 2 (14 - 2)$
$15 + 2 (12)$
$15 + 24$
39

2) $(42 - ((16 - 2^3) \div -2) \, 3) + (-2) - 34$
$(42 - ((16 - 8) \div -2) \, 3) + (-2) - 34$
$(42 - (8 \div -2) \, 3) + (-2) - 34$
$(42 - (-4) \, 3) + (-2) - 34$
$(42 - (-12)) + (-2) - 34$
$54 + (-2) - 34$
$52 - 34$
18

3) $((4 (-3 + 5) - 4^2) \div 2) + 7 (-4 + 5)$
$((4 (-3 + 5) - 16) \div 2) + 7 (-4 + 5)$
$((4 (2) - 16) \div 2) + 7 (-4 + 5)$
$((8 - 16) \div 2) + 7 (1)$
$(-8 \div 2) + 7$
$(-4) + 7$
3

4) $6(27 - (2^4 + -5)) - 10) + - (2 \cdot -5)$
$6(27 - (16 + -5)) - 10) + - (2 \cdot -5)$
$6(27 - (16 + -5)) - 10) + - (-10)$
$6(27 - 11) - 10) + (10)$
$6(16 - 10) + 10$
$6(6) + 10$
$36 + 10$
46

5) $-8(16 + (8 (-25 \div 5^2) + 9) (-7))$
$-8(16 + (8 (-25 \div 25) + 9) (-7))$
$-8(16 + (8 (-1) + 9) (-7))$
$-8(16 + ((-8 + 9) (-7))$
$-8(16 + (1) (-7))$
$-8(16 + -7)$
$-8(9)$
-72

6) $(-48 \div (55 - 49)) (5) + 8 - (-33)$
$(-48 \div (6)) (5) + 8 - (-33)$
$(-48 \div 6)) (5) + 8 - (-33)$
$-8 (5) + 8 - (-33)$
$-40 + 8 - (-33)$
$-32 - (-33)$
1

Page 64

1) $((4 + (-63 \div 3^2)) \cdot -7) - (-37 + 20)$
$((4 + (-63 \div 9)) \cdot -7) - (-37 + 20)$
$((4 + (-7)) \cdot -7) - (-37 + 20)$
$((4 + (-7)) \cdot -7) - (-17)$
$(-3 \cdot -7) - (-17)$
$21 - (-17)$
38

2) $-46 + ((-1 + 3^2)^2 - 59) + (-3 \times 7)$
$-46 + ((-1 + 9)^2 - 59) + (-3 \times 7)$
$-46 + ((-1 + 9)^2 - 59) + (-21)$
$-46 + ((8)^2 - 59) + (-21)$
$-46 + (64 - 59) + (-21)$
$-46 + 5 + (-21)$
$-41 + (-21)$
-62

3) $-115 - (104 - 69) \div -7 + (-246)$
$-115 - (35) \div -7 + (-246)$
$-115 - (-5) + (-246)$
$-110 + (-246)$
-356

4) $(((36 - 6^2) - 5) - 3)^2 - (-12 + 4)^2$
$(((36 - 36) - 5) - 3)^2 - (-12 + 4)^2$
$(((36 - 36) - 5) - 3)^2 - (-8)^2$
$((0 - 5) - 3)^2 - (64)$
$(-5 - 3)^2 - (64)$
$(-8)^2 - (64)$
$64 - (64)$
0

5) $((-19 + (4 + 3)^2) - 40)^2 (-1 + (2^3))$
$((-19 + (7)^2) - 40)^2 (-1 + 8)$
$((-19 + 49) - 40)^2 (-1 + 8)$
$((-19 + 49) - 40)^2 (7)$
$(30 - 40)^2 (7)$
$(-10)^2 (7)$
$(100) (7)$
700

6) $(175 - (5(-5 \cdot 2) + 125)) + -1000$
$(175 - (5(-10) + 125)) + -1000$
$(175 - (-50 + 125)) + -1000$
$(175 - 75) + -1000$
$100 + -1000$
-900

7) $(-166 + (722 - 547))^2 - 46 + (-37)$
$(-166 + 175)^2 - 46 + (-37)$
$(9)^2 - 46 + (-37)$
$81 - 46 + (-37)$
$35 + (-37)$
-2

8) $(270 + ((48 \div -6) + 4)^2 - 261) \div -5$
$(270 + ((-8) + 4)^2 - 261) \div -5$
$(270 + ((-4)^2 - 261) \div -5$
$(270 + (16 - 261) \div -5$
$(270 + -245) \div -5$
$(25) \div -5$
-5

Page 65

1) $-9(2^3) \div (-4(78 - (10 \times 8))) + -121$
$-9(8) \div (-4(78 - (10 \times 8))) + -121$
$-9(8) \div (-4(78 - 80)) + -121$
$-72 \div (-4(-2)) + -121$
$-72 \div 8 + -121$
$-9 + -121$
-130

2) $((84 - 100) + 4 \cdot 10) - 99) + -600$
$((84 - 100) + 40) - 99) + -600$
$(((-16) + 40) - 99) + -600$
$(24 - 99) + -600$
$-75 + -600$
-675

3) $(((21 \div -7)6 + 26)^2 - 54)^2 + -99$
$((-3)6 + 26)^2 - 54)^2 + -99$
$((-18 + 26)^2 - 54)^2 + -99$
$((8)^2 - 54)^2 + -99$
$(64 - 54)^2 + -99$
$(10)^2 + -99$
$100 + -99$
1

4) $(((-17 + 35) \div -6) (40 \div -8)) - 7 \cdot 6$
$((18 \div -6) (-5)) - 7 \cdot 6$
$(-3 (-5)) - 7 \cdot 6$
$15 - 7 \cdot 6$
$15 - 42$
-27

5) $(-19 + ((16 - 4^2) \times 3) + 15)^2 + -94$
$(-19 + ((16 - 16) \times 3) + 15)^2 + -94$
$(-19 + (0 \times 3) + 15)^2 + -94$
$(-19 + 0 + 15)^2 + -94$
$(-4)^2 + -94$
$16 + -94$
-78

6) $(-3^3 + (-35 \div (-1^5 \times 5))^2 - 20)^2 - 4$
$(-27 + (-35 \div (-1 \times 5))^2 - 20)^2 - 4$
$(-27 + (-35 \div (-5))^2 - 20)^2 - 4$
$(-27 + (7)^2 - 20)^2 - 4$
$(-27 + 49 - 20)^2 - 4$
$(22 - 20)^2 - 4$
$(2)^2 - 4$
$4 - 4$
0

7) $((264 - 185) + 16) - 24 - 2(56 \div 7)$
$(79 + 16) - 24 - 2(8)$
$95 - 24 - 2(8)$
$95 - 24 - 16$
$71 - 16$
55

8) $(-9 (-1 + 48 - 39) + 55) + 3^2 \div -9$
$(-9 (-1 + 48 - 39) + 55) + 9 \div -9$
$(-9 (47 - 39) + 55) + 9 \div -9$
$(-9 (8) + 55) + 9 \div -9$
$(-72 + 55) + 9 \div -9$
$(-17) + 9 \div -9$
$-17 + -1$
-18

Page 66

1) $(((-34 - 2) \div 2^2) + -19) - 3(2 - 4(3))$
$(-36 \div 2^2) + -19) - 3(2 - 4(3))$
$(-36 \div 4) + -19) - 3(2 - 12)$
$(-9 + -19) - 3(-10)$
$-28 - (-30)$
2

2) $((54 \div -6 - (-7)) \cdot 6) - (24 \div 3)$
$((-9 - (-7)) \cdot 6) - 8$
$(-2 \cdot 6) - 8$
$-12 - 8$
-20

3) $(((36 + 34) \div -10)^2 + -54)^2 + 3(-9)$
$((70) \div -10)^2 + -54)^2 + 3(-9)$
$((-7)^2 + -54)^2 + 3(-9)$
$(49 + -54)^2 + 3(-9)$
$(-5)^2 + 3(-9)$
$25 + 3(-9)$
$25 + -27$
-2

4) $-3(19 - (-5)^2) - (7((-32 \div 2^3) - 4))$
$-3(19 - 25) - (7((-32 \div 8) - 4))$
$-3(-6) - (7((-32 \div 8) - 4))$
$-3(-6) - (7((-4) - 4))$
$-3(-6) - (7(-8))$
$-3(-6) - (-56)$
$18 - (-56)$
74

Page 66

5) $((6(-3) + 68) \div 10) - 5(-1 - 2) \div 3$
$((-18 + 68) \div 10) - 5(-3) \div 3$
$((50) \div 10) - 5(-3) \div 3$
$((50) \div 10) - (-15) \div 3$
$5 - (-5)$
10

6) $((((-54 + -46) \div 10)4) \div 5) - 12 - 8$
$((((-100) \div 10)4) \div 5) - 12 - 8$
$(((-10)4) \div 5) - 12 - 8$
$(-40 \div 5) - 12 - 8$
$-8 - 12 - 8$
$-20 - 8$
-28

7) $(9(-48 \div 8) - 3) - 1^{10} -2(9 -15)$
$(9(-48 \div 8) - 3) - 1 -2(9 -15)$
$(9(-6) - 3) - 1 -2(-6)$
$(-54 - 3) - 1 - (-12)$
$-57 - 1 - (-12)$
$-58 - (-12)$
-46

9) $((((-13 - 15) \div 4) + 17) \div -2) - 5$
$9((-28 \div 4) + 17) \div -2) - 5$
$9((-7 + 17) \div -2) - 5$
$9(10 \div -2) - 5$
$9(-5) - 5$
$-45 - 5$
-50

Page 67

1) $((99 - 57) \div -6 + 13) - ((-2)^3 + (-1)^5)$
$(42 \div -6 + 13) - ((-2)^3 + (-1)^5)$
$(-7 + 13) - ((-2)^3 + (-1)^5)$
$6 - ((-2)^3 + (-1)^5)$
$6 - (-8 + (-1)^5)$
$6 - (-8 + -1)$
$6 - (-9)$
15

2) $-43 - (((106 + 95) - 187) \div 2) (-6)$
$-43 - ((201 - 187) \div 2) (-6)$
$-43 - (14 \div 2) (-6)$
$-43 - (7) (-6)$
$-43 - (-42)$
-1

3) $(((8 \bullet -3) \div 2 - 2^4 \bullet 0)3 + -10) + 11$
$((-24 \div 2 - 2^4 \bullet 0)3 + -10) + 11$
$((-24 \div 2 - 16 \bullet 0)3 + -10) + 11$
$((-12 - 16 \bullet 0)3 + -10) + 11$
$((-12 - 0)3 + -10) + 11$
$((-12)3 + -10) + 11$
$(-36 + -10) + 11$
$-46 + 11$
-35

4) $(2(-303 + 292) - 42) \div (-2 \bullet 3 + -2)$
$(2(-11) - 42) \div (-2 \bullet 3 + -2)$
$(-22 - 42) \div (-6 + -2)$
$(-64) \div (-6 + -2)$
$-64 \div (-8)$
8

5) $(((6(-2) \div 2^2) (-8)) \div 4) - 6 + -19$
$(((-12) \div 2^2) (-8)) \div 4) - 6 + -19$
$(((-12) \div 4) (-8)) \div 4) - 6 + -19$
$((-3) (-8)) \div 4) - 6 + -19$
$(24 \div 4) - 6 + -19$
$6 - 6 + -19$
$0 + -19$
-19

6) $((18 + 16 - 5^2)^2 - 82) + (9 + -108)$
$(34 - 25)^2 - 82) + (9 + -108)$
$(9)^2 - 82) + (-99)$
$(81 - 82) + (-99)$
$-1 + (-99)$
-100

7) $2^4(-6 - 7 - 2^2 - 1 \times 2^3) \div 5) + 4)$
$16((-6 - 7 - 4 - 1 \times 8) \div 5) + 4)$
$16((-6 - 7 - 4 - 8) \div 5) + 4)$
$16((-13 - 4 - 8) \div 5) + 4)$
$16((-17 - 8) \div 5) + 4)$
$16((-25) \div 5) + 4)$
$16(-5 + 4)$
$16(-1)$
-16

8) $(203 -128) + (2^2(8 + 3) - 24) - 56$
$(75) + (2^2(11) - 24) - 56$
$75 + (4(11) - 24) - 56$
$75 + (44 - 24) - 56$
$75 + 20 - 56$
$95 - 56$
39

Page 68

1) $-7(69 - (10 \times 8)) + 2(-505 + 497)$
$-7(69 - (80)) + 2(-505 + 497)$
$-7(69 - (80)) + 2(-8)$
$-7(- 11) + 2(-8)$
$77 + (-16)$
61

2) $-6 + -5(((83 - 9^2) \times 3 + -4)^2 - 12)$
$-6 + -5(((83 - 81) \times 3 + -4)^2 - 12)$
$-6 + -5((2 \times 3 + -4)^2 - 12)$
$-6 + -5((6 + -4)^2 - 12)$
$-6 + -5((2)^2 - 12)$
$-6 + -5(4 - 12)$
$-6 + -5(-8)$
$-6 + 40$
34

3) $((((91 - 63) \div -7)6 + 26)^2 - 3)^2$
$(((28 \div -7)6 + 26)^2 - 3)^2$
$((-4)6 + 26)^2 - 3)^2$
$((-24 + 26)^2 - 3)^2$
$((2)^2 - 3)^2$
$(4 - 3)^2$
$(1)^2$
1

4) $44 \div - 4 - 7 - 5((-89 + 71) \div -6)$
$44 \div - 4 - 7 - 5((-18) \div -6)$
$-11 - 7 - 5((-18) \div -6)$
$-11 - 7 - 5(3)$
$-11 - 7 - 15$
$-18 - 15$
-33

5) $(((177 - 194) + 3\bullet10) - 101) + -2$
$((- 17) + 3 \bullet10) - 101) + -2$
$((-17 + 30) - 101) + -2$
$(13 - 101) + -2$
$-88 + -2$
-90

6) $-27 + ((-15 \div (1^6 \times 5))^2 - 3\bullet4)^2 - 4$
$-27 + ((-15 \div (1 \times 5))^2 - 3\bullet4)^2 - 4$
$-27 + ((-15 \div 5)^2 - 3\bullet4)^2 - 4$
$-27 + ((-3)^2 - 3\bullet4)^2 - 4$
$-27 + (9 - 3\bullet4)^2 - 4$
$-27 + (9 - 12)^2 - 4$
$-27 + (-3)^2 - 4$
$-27 + 9 - 4$
$-18 - 4$
-22

7) $(2(- 74 + 66) + 48 - 115) - (3^2 \bullet-9)$
$(2(-8) + 48 - 115) - (3^2 \bullet-9)$
$(2(-8) + 48 - 115) - (9\bullet-9)$
$(-16 + 48 - 115) - (-81)$
$(32 - 115) - (-81)$
$(-83) - (-81)$
-2

8) $(9(-1 + 114 - 105) - 53 - 3^2) \div -5$
$(9(-1 + 114 - 105) - 53 - 9) \div -5$
$(9(113 - 105) - 53 - 9) \div -5$
$(9(8) - 53 - 9) \div -5$
$(72 - 53 - 9) \div -5$
$(19 - 9) \div -5$
$10 \div -5$
-2

Page 69

1) $-29 + ((78 - 24) \div 3^2) (-2) + -60$
$-29 + ((54) \div 3^2) (-2) + -60$
$-29 + (54 \div 9) (-2) + -60$
$-29 + (6) (-2) + -60$
$-29 + -12 + -60$
$-41 + -60$
-101

2) $(-5((2^2 \bullet (-3)^2) \div 6) + -11) - 5^2 + - 1$
$(-5((4 \bullet 9) \div 6) + -11) - 25 + - 1$
$(-5(36 \div 6) + -11) - 25 + - 1$
$(-5(6) + -11) - 25 + - 1$
$(-30 + -11) - 25 + - 1$
$-41 - 25 + - 1$
$-66 + - 1$
-67

3) $(3^2 - (21- 43) - 15) - 5^2 - 4 \div -2$
$(9 - (21- 43) - 15) - 25 - 4 \div -2$
$(9 - (-22) - 15) - 25 - 4 \div -2$
$(9 - (-22) - 15) - 25 - (-2)$
$(31 - 15) - 25 - (-2)$
$16 - 25 - (-2)$
$-9 - (-2)$
-7

4) $(((-6 + -2) + (-92 + 60)) \div 8) (-4)$
$((-8 + (-92 + 60)) \div 8) (-4)$
$((-8 + (-32)) \div 8) (-4)$
$(-40 \div 8) (-4)$
$(-5) (-4)$
20

5) $8((6^2 \div 9)^2 - 16)2^4 + -2(46 - 53)$
$8(36 \div 9)^2 - 16)16 + -2(46 - 53)$
$8((4)^2 - 16)16 + -2(-7)$
$8(16 - 16)16 + 14$
$8(0)16 + 14$
$0 + 14$
14

6) $2^3(500 - 603 + 96) \div 7 - 4\bullet-2$
$2^3(-103 + 96) \div 7 - 4\bullet-2$
$2^3(-7) \div 7 - 4\bullet-2$
$8(-7) \div 7 - 4\bullet-2$
$8(-7) \div 7 - (-8)$
$-56 \div 7 - (-8)$
$-8 - (-8)$
0

7) $((302 - 289) - 4^2)^2 + -1 \times 2^3 - 37$
$((13) - 4^2)^2 + -1 \times 2^3 - 37$
$(13 - 16)^2 + -1 \times 8 - 37$
$(-3)^2 + -1 \times 8 - 37$
$(-3)^2 + -8 - 37$
$9 + -8 - 37$
$1 - 37$
-36

8) $-5(993 - 1000) - 89) \div 6 + -3(6)$
$(-5(-7) - 89) \div 6 + -3(6)$
$(35 - 89) \div 6 + -18$
$-54 \div 6 + -18$
$-9 + -18$
-27

Page 70

1) $(((-47 + 23) \div 6)^2 + -72 \div 3^2)^2$
$(((-24 \div 6)^2 + -72 \div 3^2)^2$
$((-24 \div 6)^2 + -72 \div 9)^2$
$((-4)^2 + -8)^2$
$(16 + -8)^2$
$(8)^2$
64

2) $-79 + ((2 + -5)^2 - 69) - (103 + 36)$
$-79 + ((-3)^2 - 69) - (139)$
$-79 + (9 - 69) - (139)$
$-79 + - 60 - 139$
$-139 - 139$
-278

3) $(((-121 + 11^2) - 2) - 4)^2 - (12 - 4)^2$
$((-121 + 121) - 2) - 4)^2 - (12 - 4)^2$
$((0 - 2) - 4)^2 - (8)^2$
$(-2 - 4)^2 - 64$
$(-6)^2 - 64$
$36 - 64$
-28

4) $-117 - 4((-104 + 59) \div -5) + -28$
$-117 - 4((-45) \div -5) + -28$
$-117 - 4(9) + -28$
$-117 - 36 + -28$
$-153 + -28$
-181

5) $((-19 + (4 + 3)^2) - 40)^2 (-1 + (2^3))$
$((-19 + (7)^2) - 40)^2 (-1 + 8)$
$((-19 + 49) - 40)^2 (7)$
$(30 - 40)^2 (7)$
$(-10)^2 (7)$
$(100) (7)$
700

6) $(275 - (3(4 \bullet -2) - 121)) - 500$
$(275 - (3(-8) - 121)) - 500$
$(275 - (-24 - 121)) - 500$
$(275 - (-145)) - 500$
$420 - 500$
-80

7) $(171 + (725 - 900))^2 + 96 + (-10)^2$
$((171 + -175)^2 + 96 + (-10)^2$
$(171 + -175)^2 + 96 + -100$
$(-4)^2 + 96 + -100$
$16 + 96 + -100$
$112 + -100$
12

8) $(301 + ((-32 \div 4) + 1)^2 - 30) \div -8$
$(301 + ((-8) + 1)^2 - 30) \div -8$
$(301 + ((-7)^2 - 30) \div -8$
$(301 + 49 - 30) \div -8$
$(350 - 30) \div -8$
$320 \div -8$
-40

Page 71:

1) 5:8 2) 2:7 3) 6:1 4) 7:13 5) 15:11 6) 20:3 7) 2:5 8) 14:30 9) 100:9
10) 25:80 11) 6 to 7 12) 13 to 5 13) 4 to 10 14) 1 to 8 15) 60 to 100 16) 13 to 30
17) 51 to 19 18) 18 to 25 19) 7 to 5 20) 24 to 12 21) 8/3 22) 9/16 23) 25/75
24) 16/32 25) 1/8 26) 81/27 27) 6/19 28) 15/3 29) 7/18 30) 5/100

Page 72:

1) 5 to 2 2) 2 to 5 3) 7 to 3 4) 3 to 7 5) 9:1 6) 1:9 7) 5:1 8) 1:7 9) 4:3
10) 5:1 11) 3:4 12) 6:1 13) 1:4 14) 7:1 15) 5:4 16) 8:9 17) 1:9 18) 3:5 19) 1:4

Page 73:

1) R $\dfrac{25\ mg}{kg}$ 25:1   2) P $\dfrac{25\ mg}{kg} = \dfrac{1{,}000\ mg}{40\ kg}$ 25:1::1,000:40

3) P $\dfrac{13\ pigs}{6\ goats} = \dfrac{39\ pigs}{18\ goats}$ 13:6::39:18   4) R $\dfrac{39\ pigs}{18\ goats}$ 39:18

5) R $\dfrac{\$30}{hr}$ 30:1   6) P $\dfrac{\$30}{hr} = \dfrac{\$120}{4hrs}$ 30:1::120:4   7) R $\dfrac{3\ hours}{\$100}$ 3:100

8) R $\dfrac{90\ km}{hr}$ 90:1   9) P $\dfrac{90\ km}{hr} = \dfrac{450 km}{5hrs}$ 90:1::450:5

10) P $\dfrac{45\ boxes}{3\ trucks} = \dfrac{15\ boxes}{truck}$ 45:3::15:1

Page 74:
1) 12 2) 9 3) 5 4) 2 5) 144 cookies 6) 64 plates

Page 75:
1) 5 2) 6 3) 14 4) 868 5) 9 6) 7 7) 10 8) 10 9) 42
10) 96 11) 114 12) 8 13) 54 14) 8 15) 72 16) 7
17) 350 18) 3 19) 8 20) 9 21) 3 22) 15 23) 4 24) 14
25) 320 miles 26) 10 liters 27) 35 women
28) 6 ounces 29) 25 cups of flour 30) 8 gallons

Page 76:
1) 15 2) 150 3) 6 4) 76 5) 12 6) 2 7) 1 8) 10 9) 11
10) 10 11) 255 12) 722 13) 4 14) 4 15) 252 16) 1
17) 2 18) 5 19) 4 20) 2 21) 12 22) 2 23) 4 24) 6
25) 12 rows 26) 24 cups 27) 36 minutes
28) 9.14 minutes 29) 28 pizzas 30) 88 doughnuts

Page 77:
1) 90 in 2) 154 g 3) 16 ft 4) 160 lb 5) 117.5 cm 6) 95 cm
7) 15.75 kg 8) 77.78 lb 9) 40 in 10) 560 g 11) 0.75 ft
12) 120 in 13) 2 lb 14) 104 oz 15) 2.5 oz 16) 6.4 oz 17) 6 in

Page 78:
1) 103,680 s 2) 756 hr 3) 5 weeks 4) 10,080 min
5) 25,200 s 6) 3.5 days 7) 9,360 min 8) 1.6 hr
9) 60,480 s 10) 4 days 11) 3,360 hr
12) 259,200 s 13) 39,600 s 14) 2.75 weeks
15) 11,520 min 16) 3,960 min 17) 12,600 s

Page 79:
1) 4,000 mL 2) 438 pine tress 3) 1,200 lb
4) 37.5 kg 5) 25 mL 6) 270 black shirts

Page 80:
1) 10,500 lb 2) 480 g 3) 80 bales 4) 66 balls
5) 185 seeds 6) 104 red flowers 7) 135 min
8) 30 pizzas 9) 700 mL 10) 2,565 miles
11) 27 buckets 12) 105 mL

Page 81:
1) (5, 3) 2) (3, −4) 3) (−3, 7) 4) (−3, −3) 5) (−6, −3) 6) (4, 0)
7) (−8, 4) 8) (−7, −7) 9) (6, −2) 10) (1, −3) 11) (0, 0)
12) (8, −8) 13) (0, 6) 14) (7, 5) 15) (−6, 8) 16) (−1, 1)
17-36) See Chart

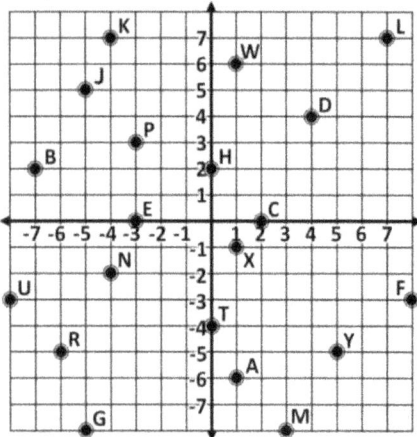

Page 82:
1) Point J 2) Point T 3) Point D 4) Point W 5) Point P
6) Point B 7) Point H 8) Point X 9) Point G 10) Point Z
11) Point R 12) Point Y 13) Point L 14) Point E 15) Point A
16) Point M 17) Point N 18) Point F 19) Point K 20) Point C
21) 4 units 22) 2 units 23) 3 units 24) 6 units 25) 2 units
26) 7 units 27) 10 units 28) 10 units 29) 4 units 30) 11 units
31) 16 units 32) 7 units 33) 4 units 34) 6 units 35) 10 units
36) 3 units 37) 14 units 38) 3 units 39) 9 units 40) 5 units

Page 83:
1) Y 2) P 3) G 4) C 5) R 6) 3 and 8 7) 11 and 2
8) 0 and 8 9) 14 and 8 10) 7 and 5
11) 2 12) 7 13) 3 14) 7 15) 8 16) 1 17) 11 18) 13
19) 18 20) 18 21) 24 22) 27 23) 38 24) 70

Page 84:
1) Point G 2) Point K 3) Point J 4) Point A
5) Point H 6) (0, 20) 7) (0.5, −15) 8) (−1.5, −25)
9) (−1.5, 0) 10) (2, 15) 11) Point X 12) Point O
13) Point U 14) Point T 15) Point R 16) Point Y
17-22 answers should be estimates.
17) (20, −50) 18) (−30, 60) 19) (2, −77) 20) (19, 80)
21) (25, 27) 22) (−25, −30) 23-40) See Chart

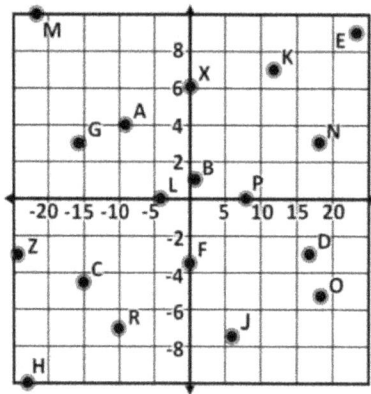

Page 85:

1)
X	Y
−3	0
−2	1
−1	2
0	3
1	4
2	5
3	6

2)
X	Y
−3	−9
−2	−6
−1	−3
0	0
1	3
2	6
3	9

3)
X	Y
−3	8
−2	7
−1	6
0	5
1	4
2	3
3	2

4)
X	Y
−3	−28
−2	−18
−1	−8
0	2
1	12
2	22
3	32

5)
X	Y
−3	−25
−2	−20
−1	−15
0	−10
1	−5
2	0
3	5

6)
X	Y
−3	−10
−2	−9
−1	−8
0	−7
1	−6
2	−5
3	−4

7)
X	Y
−3	−2
−2	−1
−1	0
0	1
1	2
2	3
3	4

8)
X	Y
−3	−6
−2	−4
−1	−2
0	0
1	2
2	4
3	6

Page 86:

1)
X	Y
−5	0
−2	3
0	5
3	8
7	12
11	16
16	21

2)
X	Y
1	2
2	1
5	−2
7	−4
9	−6
20	−17
23	−20

3)
X	Y
−30	−15
−24	−12
−14	−7
−10	−5
−2	−1
4	2
12	6

4)
X	Y
−9	19
−8	17
−5	11
0	1
3	−5
4	−7
7	−13

5)
X	Y
−10	−20
−8	−16
−7	−14
−6	−12
−4	−8
−2	−4
−1	−2

6)
X	Y
−40	−45
−30	−35
−15	−20
−5	−10
0	−5
4	−1
8	3

7)
X	Y
−50	−100
−20	−40
−15	−30
−10	−20
−3	−6
1	2
7	14

8)
X	Y
2	−6
3	−5
8	0
10	2
16	8
21	13
30	22

9)
X	Y
−90	−80
−57	−47
−32	−22
−6	4
4	14
12	22
35	45

10)
X	Y
−10	−31
−8	−25
−3	−10
−1	−4
0	−1
2	5
5	14

11)
X	Y
−15	−21
−11	−17
−6	−12
3	−3
5	−1
9	3
14	8

12)
X	Y
−5	−25
−4	−20
−1	−5
2	10
6	30
8	40
10	50

Page 87:
1)
X	Y
−4	−6
−3	−5
−1	−3
0	−2
2	0
6	4

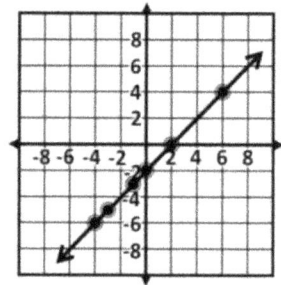

2)
X	Y
−4	10
−2	8
0	6
3	3
6	0
10	−4

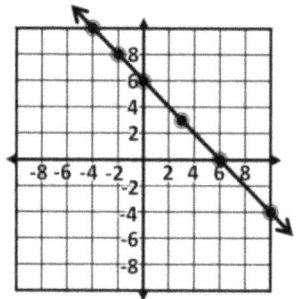

3)
X	Y
−4	−8
−2	−4
0	0
1	2
3	6
4	8

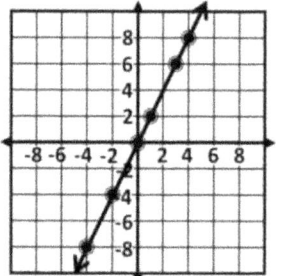

4)
X	Y
−10	−5
−8	−4
−5	−2.5
2	1
7	3.5
10	5

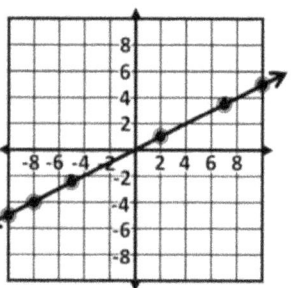

Page 88:
1) 0, H 2) Undefined, V 3) −1/4, N 4) 1/4, P
5) 3/2, P 6) −4/3, N 7) Undefined, V 8) 0, H

Page 89:
1)
X	Y
−20	−4
−10	−2
0	0
5	1
25	5

Slope = 1/5

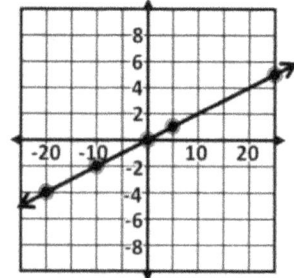

2)
X	Y
−6	−10
−4	−8
1	−3
4	0
8	4

Slope = 1/1

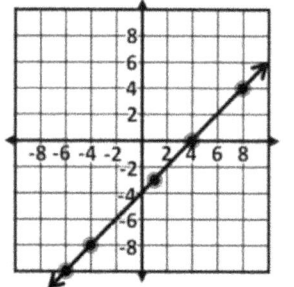

© Libro Studio LLC

110) Answer Key

Page 89 continued:

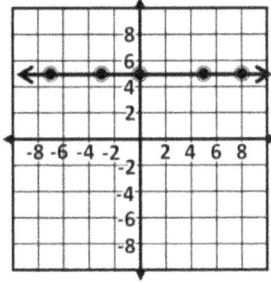

X	Y
−7	5
−3	5
0	5
5	5
8	5

Slope = 0/1

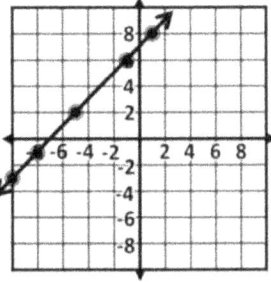

X	Y
−10	−3
−8	−1
−5	2
−1	6
1	8

Slope = 1/1

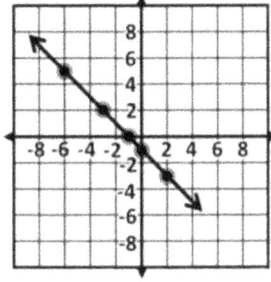

X	Y
−6	5
−3	2
−1	0
0	−1
2	−3

Slope = −1/1

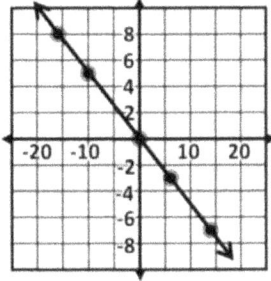

X	Y
−16	8
−10	5
0	0
6	−3
14	−7

Slope = −1/2

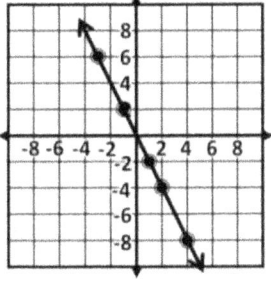

X	Y
−3	6
−1	2
1	−2
2	−4
4	−8

Slope = −2/1

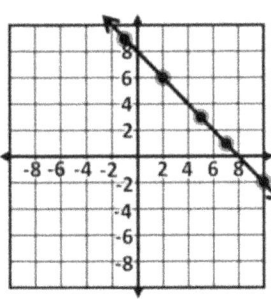

X	Y
−1	9
2	6
5	3
7	1
10	−2

Slope = −1/1

Page 90:
1) -4/7 2) 4/3 3) 8/1 4) 1/2
5) -5/2 6) -1/1 7) 2/1 8) -7/9
9) 0 10) 1/3 11) 3/1 12) 1/1
13) 5/3 14) -4/3 15) -10/1 16) undefined

Page 91:
1) 11 2) -0.14 3) -2.75 4) 47
5) 3.6 6) 3 7) 94 8) 1
9) 3 10) 8 11) 19.57 12) -3
13) 10 14) -5 15) 6 16) 52
17) 7 18) 10 19) 8 20) 100
21) -1 22) 3 23) 2 24) 5 25) 125

Page 92:
1) 2 2) 1.1 3) 87 4) 8
5) -4 6) 0.7 7) 12 8) 9
9) 10.5 10) 6.95 11) -1.8 12) 14
13) 2.5 14) -5 15) 16 16) 3
17) 3.7 18) -15 19) 4 20) 3.5
21) 50 22) -2.5 23) 3 24) 16
25) 6.35 26) 250 27) -85 28) 1250
29) 2.65

Page 93:
1) 86 2) 14 3) 50 4) 16
5) 30 6) 9.4 7) 9.7 8) 34
9) 55 10) 409 11) 220 12) 3
13) 0 and −5 14) no mode 15) −5
16) 25 17) 7 18) 12 and −14 19) 5, 36 and 82
20) 0.5 21) no mode 22) 6, 11 and 17
23) no mode

Page 94:
1) −4, −4, 5, −2 2) 20, 20, 26, no mode
3) 2.8, 7, 20, −8 4) −1.83, −1.7, 2.6, no mode
5) 5, 4, 9, no mode 6) −9.17, −6.5, 18, −4
7) 2, 2.5, 7.7, no mode 8) 36, 30, 55, 30
9) −7.2, −10, 21, −13 10) 2.5, 3, 43, −7 and 13
11) 4, 4, 4, 3 and 5 12) 10.8, 15, 37, no mode
13) 235.2, 215, 160, no mode
14) 18.5, 19.5, 6, 20
15) 36.67, 30.5, 52.3, no mode

Page 95:
1) mean: 23.3 2) range: 13 degrees
3) mode: 11, 17, 18, 19, 20
   18 occurred most often
4) mean: 16.75
 Therefore, 18 − 16.75 = 1.25 points
5) range: 27 seconds 6) median: 7:56
7) mean: $39.70 8) range: $2.66

Page 96:
1) 1, 6, 1/6 2) 1, 2, 1/2 3) 2, 6, 1/3
4) 4, 52, 1/13 5) 3, 6, 1/2 6) 8, 52, 2/13
7) 5, 6, 5/6 8) 13, 52, 1/4
9) 1, 52, 1/52 10) 39, 52, 39/52

Page 97:
1) 5/6, 0.833, 83.3% 2) 1/26, 0.038, 3.8%
3) 3/4, 0.75, 75% 4) 1/3, 0.333, 33.3%
5) 1/2, 0.5, 50% 6) 4/13, 0.308, 30.8%
7) 1/4, 0.25, 25% 8) 3/20, 0.15, 15%
9) 4/5, 0.8, 80% 10) 13/20, 0.65, 65%
11) 3/5, 0.6, 60% 12) 3/5, 0.6, 60%

Page 98:
1) 1/2 × 1/2 × 1/2 = 1/8
   1/8, 0.125, 12.5%
2) 1/2 × 1/2 = 1/4
   1/4, 0.25, 25%
3) 1/24, 0.042, 4.2%
4) 1/16, 0.063, 6.3%
5) 5/6 × 5/6 × 5/6 × 5/6 × 5/6 = 3125/7776
   3125/7776, 0.402, 40.2%
6) 1/2 × 1/6 = 1/12
   1/12, 0.083, 8.3%
7) 1/2 × 1/6 × 1/6 = 1/72
   1/72, 0.014, 1.4%
8) 2/3 × 2/3 × 2/3 = 8/27
   8/27, 0.296, 29.6%
9) 1/2 × 1/4 = 1/8
   1/8, 0.125, 12.5%
10) 1/2 × 1/2 × 2/3 = 1/6
   1/6, 0.167, 16.7%
11) 2/3 × 2/3 × 2/3 × 2/3 = 16/81
   16/81, 0.198, 19.8%
12) 1/2 × 1/13 = 1/26
   1/26, 0.038, 3.8%

Page 99:
1) 7/10 × 6/9 = 42/90 = 7/15
2) 3/10 × 2/9 × 1/8 = 6/720 = 1/120
3) 7/10 × 3/9 = 21/90 = 7/30
4) 4/52 × 3/51 = 12/2,652 = 1/221
5) 39/52 × 38/51 = 1,482/2,652 = 19/34
6) 8/52 × 7/51 × 6/50 = 336/132,600 = 14/5,525

Page 100:
1) 4/52 × 4/52 = 16/2,704 = 1/169
2) 13/52 × 12/51 × 11/50 = 1,716/132,600
   = 11/850
3) 44/52 × 44/52 × 44/52 × 44/52
   = 11/13 × 11/13 × 11/13 × 11/13
   = 14,641/28,561
4) 5/15 × 4/14 × 3/13 = 60/2,730 = 2/91
5) 2/15 × 1/14 = 2/210 = 1/105
6) 2/15 × 2/15 = 4/225
7) 8/15 × 8/15 × 8/15 = 512/3,375
8) 2/15 × 5/14 × 4/13 = 40/2,730 = 4/273
9) 8/15 × 7/14 × 6/13 × 5/12 × 4/11
   = 6,720/360,360 = 8/429
10) 8/15 × 2/15 = 16/225
11) 3/20 × 3/20 × 3/20 = 27/8,000
12) 6/20 × 5/19 × 4/18 × 3/17 = 360/116,280
   = 1/323
13) 11/20 × 10/19 = 11/38
14) 11/20 × 11/20 = 121/400
15) 11/20 × 6/19 × 5/18 = 330/6840 = 11/228
16) 6/20 × 6/20 × 6/20 × 6/20 × 6/20 × 6/20
   = 729/1,000,000
17) 6/20 × 5/19 × 4/18 × 3/17 × 2/16 × 1/15
   = 1/38,760
18) 3/20 × 11/19 = 33/380

ISBN: 978-1-63578-565-4
Current contact information can be found at:
www.UnstoppableOwl.com   www.LibroStudioLLC.com

Libro Studio LLC
9169 W State St #548
Garden City, ID 83714

Disclaimers:

This book and its contents are provided "AS IS" without warranty of any kind, expressed or implied, and hereby disclaims all implied warranties, including any warranty of merchantability and warranty of fitness for a particular purpose.

The creator and publisher DO NOT GUARANTEE THE ACCURACY, RELIABILITY, OR COMPLETENESS OF THE CONTENT OF THIS BOOK OR RELATED RESOURSES AND IS NOT RESPONSIBLE FOR ANY ERRORS OR OMISSIONS. We apologize for any inaccurate, outdated, or misleading information. Feel free to contact us if you have questions or concerns.

ALWAYS SEEK ADVICE FROM A QUALIFIED PROFESSIONAL BEFORE USING THIS BOOK OR MAKING DECISIONS BASED ON THE INFORMATION FOUND IN THIS BOOK OR RELATED RESOURCES. The creator and publisher are not liable for any decision made or action taken based on this book's content and information nor that of any related resource. You and any other persons are responsible for your own judgments, decisions, and actions.

Libro Studio LLC publishes books and other content in a variety of print and electronic formats. Some content that appears in one format may not be available in other formats. For example, some content found in a print book may not be available in the eBook format, and vice versa. Furthermore, Libro Studio LLC reserves the right to update, alter, unpublish, and/or republish the content of any of these formats at any time.

www.ingramcontent.com/pod-product-compliance
Lightning Source LLC
Chambersburg PA
CBHW051415200326
41520CB00023B/7239